Florian Baumgärtner

Atom Chip BEC Interferometer

Florian Baumgärtner

Atom Chip BEC Interferometer

Measurement of the Acceleration of Free Fall

Südwestdeutscher Verlag für Hochschulschriften

Imprint

Any brand names and product names mentioned in this book are subject to trademark, brand or patent protection and are trademarks or registered trademarks of their respective holders. The use of brand names, product names, common names, trade names, product descriptions etc. even without a particular marking in this work is in no way to be construed to mean that such names may be regarded as unrestricted in respect of trademark and brand protection legislation and could thus be used by anyone.

Cover image: www.ingimage.com

Publisher:
Südwestdeutscher Verlag für Hochschulschriften
is a trademark of
Dodo Books Indian Ocean Ltd. and OmniScriptum S.R.L publishing group

120 High Road, East Finchley, London, N2 9ED, United Kingdom
Str. Armeneasca 28/1, office 1, Chisinau MD-2012, Republic of Moldova, Europe
Printed at: see last page
ISBN: 978-3-8381-3190-0

Zugl. / Approved by: Imperial College London, PhD, 2011

Copyright © Florian Baumgärtner
Copyright © 2012 Dodo Books Indian Ocean Ltd. and OmniScriptum S.R.L publishing group

Contents

1 Introduction   5
    1.1 Coherence of Atoms 5
       1.1.1 Atom Diffraction 6
       1.1.2 Atom Interferometer 6
    1.2 Technological Advances 8
       1.2.1 Trapping of Atoms 8
       1.2.2 Cooling of Atoms 9
       1.2.3 Bose-Einstein Condensation 10
       1.2.4 Atom Chips 10
    1.3 Principle of the Experiment 11
       1.3.1 Interference Pattern 11
       1.3.2 Applying a Phase Shift 12
    1.4 Thesis Outline 13

2 Theoretical Background   15
    2.1 Introduction 15
    2.2 The Phase 16
       2.2.1 Phase of a Particle 16
       2.2.2 Phase of a BEC 17
    2.3 Theory of Matter Wave Interference 19
       2.3.1 Mean Field Picture 19
       2.3.2 Two Mode Theory 20
       2.3.3 Dephasing of a Coherent State 24
       2.3.4 Contrast in Elongated Systems 25

3 Experimental Requirements   27
    3.1 Introduction 27
    3.2 Infrastructure 27
       3.2.1 Vacuum System 27
       3.2.2 The Atom Chip 29
       3.2.3 Laser System 31

		3.2.4	Imaging System	34
		3.2.5	External Magnetic Fields	35
		3.2.6	Current Control	36
		3.2.7	Radio-Frequency Coupling	38
	3.3	Making BECs		39
		3.3.1	Computer Control	39
		3.3.2	Experiment Cycle	40
4	**Characterising the Magnetic Wire Trap**			**42**
	4.1	The Magnetic Wire Trap		42
		4.1.1	Ioffe-Pritchard trap	43
		4.1.2	Magnetic Trapping Fields	44
	4.2	Characterisation of the BEC		45
		4.2.1	BEC near the Chip Surface	45
		4.2.2	Trap Frequencies	47
		4.2.3	Trap Bottom	48
		4.2.4	Trap Lifetime	49
	4.3	Calibration of the Imaging System		50
		4.3.1	Magnification from Gravity	50
		4.3.2	Calibration with Magnetic Fields	51
5	**Symmetric and Asymmetric Double Well Potential**			**53**
	5.1	Introduction		53
	5.2	RF-Adiabatic Potentials		54
		5.2.1	The dressed state Hamiltonian	54
		5.2.2	The Rotating Wave Approximation	56
		5.2.3	Double Well Spectroscopy	57
	5.3	Splitting of the BEC		60
		5.3.1	The Double Well Potential	61
		5.3.2	The Splitting Process	62
		5.3.3	Balancing of the Double Well	62
	5.4	Implementation and Characterisation of the Asymmetric Double Well		64
		5.4.1	Tilting of the Double Well	65
		5.4.2	Gravitational Potential Difference	68
		5.4.3	Spectroscopy	70
	5.5	Conclusion		72
6	**Making the Interferometer Work**			**75**
	6.1	Introduction		75
	6.2	Matter Wave Interference		75

		6.2.1	Observation of Interference	76
		6.2.2	Extracting Information	78
	6.3	Spin States		79
		6.3.1	Projection	79
		6.3.2	Observing Spin States	81
	6.4	Phase Coherence and Stability		81
		6.4.1	Phase Distribution	82
		6.4.2	Comparison	85
	6.5	Limits of the System		87
		6.5.1	Phase Spreading	87
		6.5.2	Fringe Contrast	90
	6.6	Conclusion		92

7 Running the Interferometer — 93

    7.1 Schemes for a Gravitational Gradient Interferometer 93
        7.1.1 Symmetric Preparation 94
        7.1.2 Asymmetric Preparation 96
    7.2 The Gravitational Gradient Interferometer 98
        7.2.1 Measurement of the Balancing 99
        7.2.2 Measurement of the Acceleration of Free Fall 100
    7.3 The Asymmetric Interferometer 101
        7.3.1 Influence of the Splitting Process 101
        7.3.2 Measurement of the Coupling Length 102
    7.4 Conclusion and Error Discussion 103

8 Conclusions and Outlook — 106

    8.1 Summary 106
    8.2 Outlook 107

A Rubidium — 109

B Directional Statistics — 111

    B.1 The Mean Direction 111
    B.2 Circular Variance and Standard Deviation 112
    B.3 Distribution Functions 112

Chapter 1
Introduction

This book is thought of as a manual on how to build an atom Bose-Einstein condensate (BEC) interferometer. The starting point is a single BEC trapped near an atom chip surface. From there on we will describe the various steps to implement an interferometer and making it work. The demonstration of an absolute measurement of the acceleration of free fall will prove the capability of the device. The determination of the limitations of an atom BEC interferometer will be derived from the detailed discussion of the results and are the important conclusions of this thesis.

Everybody should be warned in advance that at the end of our journey we will not end up with a small and handy device. The working interferometer still takes a whole optical table and has the complexity of most typical ultra-cold atom experiments. However, it illustrates how the principles of quantum mechanics and the concept of BEC can be used for an actual application. So far we should be allowed to draw some analogue to Konrad Zuses Z3 which is considered the world's first operational computer [1]. The Z3 demonstrated binary arithmetic and programmability while its dimension basically took a whole room. Today every mobile phone is more powerful than the Z3 and therefore a comparison with modern computers is difficult.

Such a process we call evolution and if we have a look back we will notice that also interferometry with atoms has developed a lot since the first diffraction experiments in the last century.

1.1 Coherence of Atoms

The development of the first light interferometers in the 19th century had an huge impact on the physical world view. For example, it was the Michelson-Morley experiment [2] which gave evidence that a luminiferous aether does not exist and therefore opened the way for special relativity. Today the laser provides the optimal tool for light interferometry producing perfectly coherent light with high coherence lengths. However, photons have the big drawback that they interact only very weakly with gravitational, magnetic and electric fields and therefore their use as a sensor is limited. The possibility to sense such fields with atom interferometers developed within the last decades.

1.1.1 Atom Diffraction

At the beginning of the 20th century three crucial observations provoked a discussion on the nature of light. Although its wave property was proved in Young's double slit experiment, the measurement of the spectrum of black body radiation, the photoelectric effect and the Compton scattering of light on electrons point to a particle character of light. The explanation of the photoelectric effect where a single photon transfers its whole energy to one single electron brought Albert Einstein the Noble Prize in physics in 1921. Thinking about the wave-particle dualism of light Louis de Broglie in 1924 suggested the same concept to solid particles such as neutrons, electrons and atoms. From the relation between wavelength and momentum p he derived the de Broglie wavelength of a matter wave $\lambda_{dB} = h/p$ with h being the Planck constant. Indeed only a few years later the diffraction of electrons shot through a thin crystal foil was observed [3]. Esterman and Stern [4] finally proved the wave propagation of atoms by diffraction of He on a LiF crystal surface. In both experiments the measured atom distribution showed patterns similar to the diffraction patterns of X-rays.

The diffraction of atoms was repeatedly demonstrated on various types of gratings. Advances in the fabrication processes opened the path for transmission gratings on a nm scale necessary to observe diffraction of atoms with de Broglie wavelengths of only a few pm [5, 6]. Another experiment makes use of two counter-propagating laser beams [7] with wavelength λ_{laser}. The resulting standing wave then acts as a transmission grating for He atoms with a period $\lambda_{laser}/2$. Constructive interference of neighbouring matter waves spreading from the grating occurs if the particles' path difference is a multiple of the de Broglie wavelength. Therefore we define the angles θ_n with $n = 0, 1, 2, ...$ of maximum particle density on an observation screen as

$$n\lambda_{dB} = \frac{\lambda_{laser}}{2} \sin \theta_n \tag{1.1}$$

which directly depends on the de Broglie wavelength λ_{dB}. This experiment clearly demonstrates the concept of the particle-wave duality that particles can interfere in the same way light does. Diffraction of particle projectiles is now widely used to investigate the structure of targets such as biomolecules or even other particles.

1.1.2 Atom Interferometer

The wave properties of solid particles inspired Heisenberg and Schrödinger to the idea of describing particles as wavepackets and to the development of quantum theory. Within this theory a particle is described by a wavefunction Ψ which can contain information about both external and internal states. The interesting point in terms of interferometry is that the wavefunction allows the occurance of a phase term $e^{i\phi}$ without affecting the outcome of a classical measurement. In particular the measurement outcome of a classical variable represented by the operator $\hat{O}$ is described by $|\langle \Psi^\dagger | \hat{O} | \Psi \rangle|^2$. Modifying the wavefunction with an additional phase such that $\Psi' = \Psi e^{i\phi}$ does not alter the observation of the classical variable $|\langle \Psi'^\dagger | \hat{O} | \Psi' \rangle|^2 = |\langle \Psi^\dagger | \hat{O} | \Psi \rangle|^2$. The situation, however, looks completely different if we ask for the interference pattern of two particles with the wavefunctions Ψ_a and $\Psi_b e^{i\phi}$.

Then
$$|\Psi_a + \Psi_b e^{i\phi}|^2 = |\Psi_a|^2 + |\Psi_b|^2 + 2\Psi_a^*\Psi_b \cos\phi, \qquad (1.2)$$
where the phase obviously manifests as a modulation of the total density. Equation (1.2) itself represents the quantum mechanical basis of a solid particle interferometer and tells us what is necessary to implement such a device.

First, the interferometer needs to have two arms a and b which are represented by the two wavefunctions. Second, a phase shift can be applied on one of the arms but the phase has to be well defined. Otherwise, a repeated measurement would not deliver the same result. We refer to this requirement usually also as coherence of the wavefunctions.

From these simple considerations we can define five steps which all interferometers have in common: (1) state preparation, (2) coherent splitting, (3) phase evolution, (4) coherent recombination and (5) read out of the phase. Different types of interferometers differ in the particles, the method of manipulation and the states which are used.

One early type of interferometer is the 3-grating interferometer which represents the analogue to the Mach-Zehnder setup [8] for light. The first grating separates the incoming particles by diffraction into two beam paths. The particles are recombined in the plane of the third grating after which a detector counts the particle flux in the different diffraction orders. The Mach-Zehnder interferometer separates the particles spatially and interactions between the arms are disabled. In the past this setup was successfully implemented for electrons [9], neutrons [10] and atoms [11].

Atoms with their rich structure of internal states open the path for interferometers where atoms in the two interferometer arms occupy different atomic states or superpositions of states. A high precision measurement of the acceleration of free fall was demonstrated with such a device by Peters *et al.* [12]. The interferometer achieved an absolute uncertainty $\Delta g/g = 10^{-9}$ by preparing atoms from an atomic fountain in a superposition of momentum and atomic level entangled state. The relative phase is then proportional to the gravitational field g and quadratic in the interaction time between the $\pi/2$ Raman pulses driving the atomic transitions.

In the last decade a completely new concept of atom interferometry has been developed. Instead of using freely moving atom beams these new devices manipulate the states of atoms within a confined potential. The preparation of the atoms as a BEC which exhibits long range order guarantees the coherence between the atoms. Interferometry by splitting and recombining a BEC into a double well potential was demonstrated by various groups [13, 14]. A phase shift applied to one of the two spatially separated wells is extracted from an interference pattern. The sensitivity of such a BEC interferometer is typically limited by the dephasing effects causing an increased spread in the phase measurements. A promising method to increase the coherence time seems to be the implementation of number squeezing which was reported by G.-B. Jo *et al.* [15].

Despite a lot of progress, however, an absolute measurement of an external field with confined atoms has not been performed yet. This is exactly the point where this thesis contributes to the interesting field of atomic physics.

1.2 Technological Advances

Atom interferometers based on trapped atoms only developed recently. In the following we give a brief overview over the recent key technologies and discoveries essential to build an atom chip BEC interferometer.

1.2.1 Trapping of Atoms

One important property of atoms is their interaction with light, magnetic and electric fields. The influence of laser light with a frequency near a transition of an atomic state shifts the energy level of the atom [16]. The magnitude of the shift is proportional to the intensity of the light field and depends on the sign of the detuning. Therefore an atom in a red-detuned Gaussian laser beam is accelerated towards the focus of the beam. Such traps were successfully implemented in cold atom experiments [17, 18, 19].

A similar effect is caused by a magnetic field **B**. The total atomic spin $\hat{\mathbf{F}}$ interacts with the external field and shifts the energy of the atomic state which is known as the so-called Zeeman-shift and is described by the Hamiltonian [20]

$$\hat{H}_{\text{Zeeman}} = g_F \mu_B \hat{\mathbf{F}} \cdot \hat{\mathbf{B}}(\mathbf{r}). \tag{1.3}$$

Here g_F describes the hyperfine Landé factor and μ_B is the Bohr magneton. For hydrogen like atoms in their ground state the atomic spin $\hat{\mathbf{F}}$ can be identified with the orbital angular momentum of the atom. In general the coupling of the hyperfine states with external, magnetic fields behaves non-linearly. However, most cold atom experiments operate at field strengths low enough that a linear approximation is justified

$$\Delta E_{\text{Zeeman}} = g_F \mu_B m_F |\mathbf{B}|. \tag{1.4}$$

The projection m_F of the total atomic spin onto the magnetic field takes the values $-F \leq m_F \leq F$ and defines the sign of the energy shift. If the shift is positive we end up with a low-field seeking state, while a negative sign causes the atom to be sucked into an area of higher magnetic field. The Earnshaw theorem forbids a local maximum of the absolute value of a magnetic field in free space. Therefore static magnetic traps for high-field seeking states are impossible to implement in contrast to blue-detuned optical dipole traps.

A famous example of a magnetic trap with a local, magnetic field minimum is the Ioffe-Pritchard (IP) trap. It is today a typical trapping configuration in atom chip experiments where it is created by overlapping the magnetic field of a chip wire with a homogenous external magnetic field. The magnetic field minimum is non-zero to prevent losses of trapped atoms by Landau-Zener transitions. Typically the trapping potential of an IP trap can be written in the form

$$V(r,z) = \frac{1}{2} m \omega_r^2 r^2 + \frac{1}{2} m \omega_z^2 z^2 \tag{1.5}$$

which has axial symmetry with the radial trap frequency ω_r and the longitudinal trap frequency ω_z.

The versatility of such traps is greatly enhanced by the implementation of radio-frequency (rf) adibatic potentials which were theoretically proposed by Zobay and Garraway [21, 22]. An rf magnetic field dresses the hyperfine states of the atoms. The new state of the atoms is then described by a superposition of the original states. Depending on the application of the dressing fields the atoms experience a potential of different shape. The most common implemented setup is the double well potential used to split a BEC into two wells [13, 23]. However, other trap configurations [24] are also possible.

1.2.2 Cooling of Atoms

Atoms cannot be held in a trap unless their thermal energy is substantially lower than the depth of the trap. Laser cooling in a magneto-optical trap (MOT) [25] is the standard method of achieving this. A MOT consists of a magnetic quadrupole field combined with six confining laser beams propagating from the six directions $\pm x$, $\pm y$ and $\pm z$. Atoms enter the cooling area from a background gas vapour. The laser beams are slightly detuned from the atomic cooling transition. Whenever an atom moves in a certain direction the Doppler shift causes resonance with the corresponding, counterpropagating laser beam. The atom scatters photons and experiences a net kick into the direction opposite of its movements. Each scattered photon takes a part of the kinetic energy away and the atom cools down in momentum space. The Zeeman splitting in the quadrupole field and the opposite helicity of counterpropating laser beams ensures that the atoms only experience forces towards the centre of the trap. Therefore the quadrupole field provides trapping in position space.

Laser cooling though is not possible with every kind of atom. In fact it is limited to only a few atoms of the periodic table, most of them the alkali metals, since the atomic structure has to be simple enough to implement a closed cooling cycle of internal states. Otherwise the atom can end up in a state which is not resonant to the laser beams after the first scattering process preventing further reduction of kinetic energy. As cooling occurs due to a scattering process there is a lower limit in achievable temperature T_{min} set by the natural linewidth Γ [26]

$$k_B T_{min} = \hbar \Gamma / 2 \tag{1.6}$$

with k_B the Boltzmann constant and $\hbar$ the Planck constant divided by 2π.

In order to reach even lower temperatures further cooling techniques were demonstrated in the last 20 years, i.e. Sisyphus cooling, Raman cooling and resolved-sideband Raman cooling. For neutral atoms in magnetic traps BEC transition, however, was first achieved with evaporative cooling [27, 28] where an applied rf field drives spin flips of hot atoms to untrapped m_F states. The hot atoms are expelled from the trap and take a part of the thermal energy with them. Like in a hot cup of coffee where the steam takes energy away from the liquid which cools down, the remaining atoms establish a new equilibrium at a lower temperature. By subsequently lowering the rf frequency and hence moving the resonant sphere closer to the trap centre temperatures on the order of only a few nK can be achieved in the case of large initial atom numbers.

1.2.3 Bose-Einstein Condensation

Regarding a many-body system with N particles, a BEC represents a system with a number of particles occupying a certain state which is on the order of N. Such a behaviour can be achieved in tight traps with high atomic densities. Reducing the temperature of the system decreases the selection of available states, since then most of the atoms do not have enough energy to occupy higher states.

Concerning the distribution of the atoms among the states the quantum statistical properties of bosons are crucial for condensation to occur. Quantum theory gives only a probability to find an atom at a certain point in space, hence identical atoms moving in the same volume are indistinguishable in a measurement. Assuming two states occupied by two non-identical atoms, e.g. a blue and a red atom, we will find four possible configurations to distribute the atoms among the states. In the case of indistinguishable particles, however, the number of configurations reduces to three possibilities. In fact the configurations where exactly one atom occupies each state remain unaffected after a particle swap and hence represent the same physical situation.

Taking this theoretical considerations on bosons and quantum statistics into account A. Einstein derived the equation for the transition temperature to BEC [29]

$$T_c = \frac{\hbar \omega_{\text{geo}}}{k_B} \left(\frac{N}{\zeta(3)} \right)^{1/3} \tag{1.7}$$

for non-interacting bosons in 1925. In the equation we find the Riemann ζ function and the geometric average of the trap frequencies $\omega_{\text{geo}} = \left(\omega_x \omega_y \omega_z \right)^{1/3}$. Although the existence of BEC [30] was theoretically imaginable, it took several decades until the necessary techniques for atom trapping and cooling were available. Since the first realisation of BEC in 1995, scientific interest exploded.

The complexity of BEC is extended by atom-atom interactions which play an important role in the behaviour of the system and are the focus of various studies [31, 32, 33]. The long range order within a BEC causes the coherence of the atoms which led to the demonstration of atom lasers [34, 35] in the past.

1.2.4 Atom Chips

In order to gain large trapping confinement, high field gradients are necessary. The magnetic field gradient of a single wire scales as $\partial |\mathbf{B}|/\partial r \propto 1/r^2$ which becomes large in the limit $r \to 0$. It is therefore advantageous to create a magnetic trap near a single wire instead with fields created with structures outside the vacuum chamber.

Based on this consideration tight traps with wire diameters down to $50\,\mu\text{m}$ were realised [36, 37]. Further miniaturisation finally led to the development of so-called atom chips. An atom chip consists of wires or a permanent magnetic structure placed onto a substrate. The production of atom chips enormously benefits from the well-known micro-chip technologies, and as the most important fabrication techniques electroplating and evaporated microstructures have been established. The first atom chips were created by electroplating a conductor on a sapphire substrate [38, 39]. The rough surface of electroplated structures, however, has consequencs for the smoothness of magnetic potentials near the chip surface [40]. Evaporation of thin metal films onto a Si or GaAs substrates achieved better results [41]. Since these materials are semiconductors, the substrates are insulated with a thin SiO_2 layer with a thickness up to $5\,\mu$m. The wires are patterned in a conductive surface deposited on top and have dimensions in the order of only a few μm. As a conductor a material with high reflectivity, typically gold, is processed. The reflective surface allows the implementation of a mirror-MOT which is necessary to load the magnetic trap. In recent years also the fabrication of multi-layer chips was demonstrated [42].

Depending on the exact configuration tight magnetic traps which also allow the confinement of a BEC [43, 44] or arrays of traps [45, 46] can be realised near the chip surface. Another advantage of an atom chip is the possibility of integrating micro-optical elements into the chip, e.g. pyramid-MOTs [47], waveguides [48], cavities and optical fibres for atom detection [49, 50, 51, 52].

Typical trapping distances from the surface are of the order of $10\,\mu$m to a few $100\,\mu$m. At very small distances $< 100\,\mu$m the potential suffers distortion from fragmentation and the Casimir-Polder force. Atoms depositing on the chip over time can also cause the patch effect and limit the trap depth [20]. The crucial limiting factor in the design, however, is the transport of heat away from the chip surface. Due to the small wire cross-section a high amount of power is dissipated at fairly low currents of around 1 A. If the heat conductivity of the chip is not sufficient, the wires will melt and be destroyed. At this point also the mount of the chip serving as a heat sink plays an important role.

1.3 Principle of the Experiment

In the following we give a brief overview of the principles of the experiment which should help to understand the context in the detailed discussions of the later chapters.

1.3.1 Interference Pattern

Similar to the famous double slit experiment the two BECs in a double well potential can be considered as point like sources. When we turn off the confining potential, the atom-atom interaction potential transforms into kinetic energy and the whole cloud spreads like the wave spreads from the impact area of a stone in a pond. By starting with two BECs close to each other we can overlap the expanding atoms after release. The result is not, as one would expect classically, one big atomic gas cloud. In fact ultra-cold atoms behave like a wave and we find a pattern analogue to a standing wave with areas of high and low atomic density, so-called interference fringes described by equation (1.2). A simulation of the experiment is compared with the observed interference pattern in figure 1.1.

The important observation relevant to our experiment is that the position where the fringes occur depends on the relative phase of the two BECs. The relative phase is the most interesting quantity for an interferometer and therefore a great part of this thesis deals with the observation of the interference

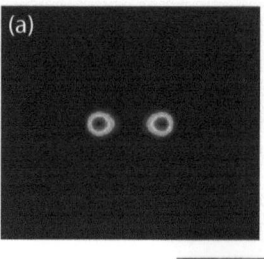 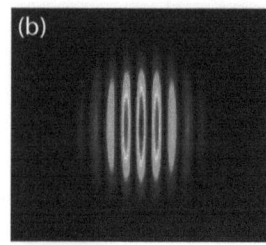
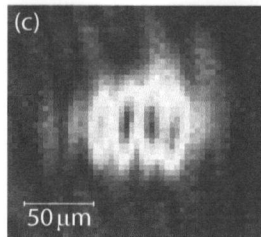

Figure 1.1: (a) Simulation by Robin Scott of two BECs confined inside two wells. After the confinement is turned off the atoms spread from the two wells and overlap (b) showing an interference pattern. (c) The interference pattern actually observed in the experiment.

pattern, the extraction of the phase and the limits of a phase measurement.

In the first experiment of this kind by Andrews *et al.* [53], the interference fringe position was not reproducible in repeated experiments due to a random relative phase between the halves. Following experiments [14, 13] finally showed coherence of the process and a repeatable relative phase measurement was guaranteed.

1.3.2 Applying a Phase Shift

Observing an interference pattern and extracting a relative phase does not yet make an interferometer. We rather have to implement a complete scheme which also allows the evolution of the phase. In general the evolution of the relative phase ϕ is given by

$$\phi = -\frac{1}{\hbar} \int_t \Delta V(\mathbf{x}) \, d\tau, \qquad (1.8)$$

and depends on the trapping time t of the atomic clouds and the potential difference ΔV between the two BECs.

If ΔV is caused by an external field, its gradient can be extracted from the interference pattern. We are interested in measuring the gravitational acceleration g and therefore it stands to reason to introduce a height difference between the two BECs.

The experimental scheme for our atom chip BEC interferometer is shown in figure 1.2. After preparing the BEC in a magnetic trap we split it coherently into a double well potential. We then introduce a potential difference between the two wells of the potential. After moving the BECs back to their initial position we recombine the clouds in free fall and observe the interference pattern.

This experimental cycle completes the five basic steps of an interferometer. Since the applied potential shift depends on various parts, a detailed analysis of the system is necessary to extract the gravitational constant g.

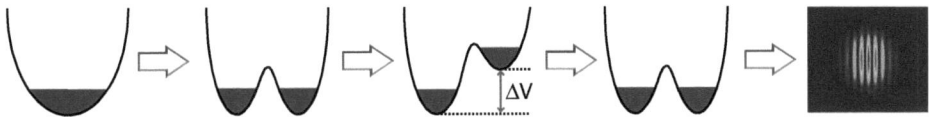

Figure 1.2: Basic procedure of the experiment. After preparation of a BEC in a magnetic trap we split it into a double well potential. By applying a potential difference the relative phase between the two condensates starts to evolve. We read out the resulting phase shift from the interference pattern after overlapping the atoms in free fall. The applied potential gradient can be determined from the phase evolution allowing the measurement of external fields.

1.4 Thesis Outline

The plan for this thesis is the following:

In chapter 2 we discuss the basic theory of matter wave interference. We examine the meaning of the phase and the origin of the interference pattern. We also describe theoretically dephasing and longitudinal phase fluctuations which limit the precision of an interferometer. The knowledge of this part is essential to understand the behaviour and the operation method of the atom BEC interferometer. A physicist from the field might be confident with the presented concepts and can consider this chapter as a reference guide.

We then move on to the experiment itself and present the needed hardware, the setup and computer control in chapter 3. Here we also describe how the experimental cycle looks like and how we achieve BEC in a magnetic trap near the chip surface.

We explore the characteristics and properties of the magnetically trapped BEC in chapter 4 which represents the starting point for our interferometer. First, we describe the magnetic trap and calculate the resulting trapping fields. Then we determine properties such as trap bottom, lifetime and trap frequencies of the confining potential from experiments. A crucial part of this chapter is also the calibration of the imaging system which is essential to extract numbers from images.

In chapter 5 we present how we split the BEC into two interferometer arms. An applied radio-frequency (rf) field dresses the atoms which then experience a double well potential. In this context we also present the method to apply an energy shift between the wells making the potential asymmetric. A detailed characterization of the asymmetric double well potential follows. The results of this chapter are crucial for a comparison with the interferometer measurements presented later in the text.

We present the observation of an interference pattern after recombining the BECs from two wells during free fall in chapter 6. We show how a phase can be extracted from the fringe pattern and that the repeated experiment delivers a reproducible relative phase. The width of the phase distribution is influenced by the turn off and the time the atoms spend inside the double well potential. In that context we explore the limits of our system due to dephasing.

Finally, in chapter 7 we combine all the steps in one single scheme which enables a working interferometer. After discussing the theory and expectations for the outcome of the measurement we run the experiment and derive the gravitational acceleration g from the shift of the relative phase with a statistical error of around 5%. In addition we present a modified scheme from which we gain additional knowledge about the double well system.

The thesis ends with our conclusions on the experimental results and an error discussion on the g measurement in chapter 8. In a short outlook we present possible improvements of the atom BEC interferometer for the future.

Chapter 2

Theoretical Background

2.1 Introduction

The concept of Louis de Broglie to assign a wavelength to particles such as electrons, neutrons and atoms and the development of quantum mechanics led to the first interference experiments for particles. The observation of an interference pattern proves the wave character of massive objects and therewith also the existence of a phase. Unlike the velocity or the mass, the phase of a particle has no physical significance and manifests itself only in interference experiments. It turns out, however, that the phase evolves by the coupling of a particle to external fields and potentials. Measuring the relative phase between two particles moving in a known and unknown potential therefore enables us to draw a conclusion on the unknown one. Since atoms couple to gravitational, magnetic and electric fields, the behaviour of their phase is a quantitiy of great interest.

The relative phase can be determined from interference experiments but mostly requires a source of coherent atoms. For an ensemble of atoms coherence is achieved by reducing momentum and its uncertainty. In the concept of BEC an accumulation of many particles occupies one single ground state where the whole ensemble has one global phase. Therefore a BEC represents an ideal source of coherent particles.

In this section we cover the basic theory needed to understand an atom BEC interferometer. We start in section 2.2 with the examination of a particle's phase shift. We then extend the picture from a single atom to a BEC where all atoms occupy the same state and can be described by an order parameter. This concept of a condensate wavefunction implies the establishment of a global phase. Two BECs overlapping in free fall show the existence of the global phase in a characteristic interference pattern. We discuss the occurrance of matter wave interference and its properties in section 2.3. The theoretical background presented in this chapter builds the basis for an atom BEC interferometer and is therefore essential to understand the procedures throughout this thesis.

2.2 The Phase

2.2.1 Phase of a Particle

We start our considerations with a single, nonrelativistic particle of mass m moving with momentum **p** in an external potential $V(\mathbf{x})$. The Hamiltonian for this simple system is written as

$$\hat{H} = \frac{\hat{\mathbf{p}}^2}{2m} + \hat{V}(\mathbf{x}). \tag{2.1}$$

Suppose we want to examine the event that the particle travels from one place $(\mathbf{x}_a, t_a)$ to another $(\mathbf{x}_b, t_b)$. The questions that arise then are: Which path will the particle take and what is its probability to happen? In the formalism of the canonical Hamiltonian the amplitude is simply given by the expression $\langle \mathbf{x}_b, t_b | e^{-i\hat{H}(t_b - t_a)/\hbar} | \mathbf{x}_a, t_a \rangle$.

Another solution for the problem can be derived from Feynman's path integral formalism [54] which is based on the superposition principle of quantum mechanics. The amplitude for the particle to travel from $(\mathbf{x}_a, t_a)$ to $(\mathbf{x}_b, t_b)$ is then nothing other than the coherent sum of the amplitudes for all possible paths. The paths are all equally important but can differ from each other by a phase ϕ_p which has no influence on the probability of a certain path $x(t)$. Hence we write the total amplitude as the functional integral [55]

$$A(\mathbf{x}_a, \mathbf{x}_b, t_a, t_b) = \int Dx(t)\, e^{i\phi_p[x(t)]} \tag{2.2}$$

where the symbol $\int Dx(t)$ means the sum over all possible paths. In the classical case, however, the path of the particle is set by the initial conditions. The classical path $x_{cl}(t)$ is unique and satisfies the principle of least action

$$\frac{\delta}{\delta x(t)} S[x(t)]|_{x_{cl}} = 0 \tag{2.3}$$

where $S = \int L(\dot{\mathbf{x}}, \mathbf{x})\, dt$ is the classical action. In order to describe nature the different theories have to be consistent with each other. Therefore we expect that in the classical limit only the classical path contributes to the amplitude. A particular path in equation (2.2) is then unique if its phase is also unique which is in fact nothing else than the classical action $S/\hbar$.

Assuming a time-independent potential $V(\mathbf{x})$ the particle with the total energy E acquires a phase shift of

$$\phi_p = \int \left[\sqrt{\frac{2m}{\hbar^2}(E - V(\mathbf{x}))} - \sqrt{\frac{2m}{\hbar^2}E}\right] d^3x. \tag{2.4}$$

Expanding this expression to first order in V/E leads us to the approximation for this phase shift of

$$\phi_p \approx -\frac{1}{\hbar v}\int V(\mathbf{x})\, dx = -\frac{1}{\hbar}\int_t V(\mathbf{x})\, d\tau, \tag{2.5}$$

where v is the velocity of the particle and $t = t_b - t_a$ is the interaction time. The phase evolution depends on both the potential and the time the particle spends inside the potential. A classical force which can be viewed as a potential gradient $\mathbf{F} = -\nabla V(\mathbf{x})$ will cause the same effect. But such a force will not only change the phase but will also change the motion of the particle. For the sake of completeness we mention that there are also topological phase shifts induced between two particles such as the Aharonov-Bohm effect [56]. Such topological effects are independent of the velocity and the path of an atom. A discussion of these is beyond the scope of this thesis.

2.2.2 Phase of a BEC

Order Parameter

In the case of a dilute Bose gas trapped in an external potential $V_{ext}(\mathbf{r}, t)$ where the s-wave scattering length is much smaller than the average distance between the atoms, Gross [57] and Pitaevskii [58] derived independently the Gross-Pitaevski (GP) equation

$$i\hbar \frac{\partial \Psi(\mathbf{r}, t)}{\partial t} = -\frac{\hbar}{2m} \nabla^2 \Psi(\mathbf{r}, t) + V_{ext}(\mathbf{r}, t) \Psi(\mathbf{r}, t) + g |\Psi(\mathbf{r}, t)|^2 \Psi(\mathbf{r}, t). \tag{2.6}$$

The GP equation is valid in the case of a large number of atoms in the condensate N_0 and for low temperatures. The last term describes atom-atom interactions with the coupling constant

$$g = \frac{4\pi \hbar^2 a}{m} \tag{2.7}$$

where m is the atomic mass and a is the s-wave scattering length. A posititive a corresponds to repulsive forces, while $a < 0$ describes an attractive interaction. For $a = 0$ we get vanishing interactions and the GP equation becomes the well-known Schrödinger equation. The function $\Psi(\mathbf{r})$ is a macroscopic Schrödinger wavefunction normalised to the number of atoms N which describes the whole ensemble of atoms within a mean-field approximation. It is called the order parameter and can be written as

$$\Psi(\mathbf{r}, t) = \sqrt{n(\mathbf{r}, t)} e^{i\phi(\mathbf{r}, t)}. \tag{2.8}$$

The order parameter has a well-defined phase ϕ and the atom density n is fixed by the relation $n(\mathbf{r}, t) = |\Psi(\mathbf{r}, t)|^2$. The construction of the classical field $\Psi(\mathbf{r})$ goes back to the ideas of Bogoliubov [59] on dilute Bose gases in the limit $N \to \infty$.

Velocity Field

In order to proceed with our examination of the order parameter and the meaning of its phase we make use of the tools provided by classical field theory. A closer look at the Hamiltonian of equation (2.6) shows that it is invariant under a transformation $\Psi \to \Psi e^{i\phi}$ which is nothing else than a phase shift of the order parameter. According to Noether's theorem every transformation that leaves the equations of motions unchanged induces a symmetry. In the case of BEC the conserved quantity is the four-current [55]

$$j^\mu = \frac{\hbar}{2im} [(\partial^\mu \Psi^*) \Psi - \Psi^* (\partial^\mu \Psi)] \tag{2.9}$$

which is a vector containing both charge and current. Combining this relation with the order parameter of equation (2.8) and taking the condition for conservation $\partial_\mu j^\mu = 0$ into account, leads us to the expression

$$\frac{\partial}{\partial t} n + \frac{\hbar}{m} \nabla (n \nabla \phi) = 0. \qquad (2.10)$$

Equation (2.10) is nothing else than the well-known equation of continuity which expresses the conservation of particles. A comparison with the classical equation of continuity suggests to define the superfluid velocity as

$$\mathbf{v}(\mathbf{r}, t) = \frac{\hbar}{m} \nabla \phi(\mathbf{r}, t). \qquad (2.11)$$

The phase of a BEC fixes its velocity field. A trapped BEC without vortices resting in the steady state has a vanishing velocity field and hence according to equation (2.11) the phase is constant over the whole condensate. Since a BEC is defined by the macroscopic occupation of the lowest energy state, the intuitive expectation is indeed that the atoms establish a global phase. A manipulation on the external potential $V_{\text{ext}}(\mathbf{r})$ is the same for each particle in the condensate experiencing a total shift in phase according to equation (2.5).

Globality

As a trapped Bose gas is cooled down below the critical temperature for condensation T_c, density and phase fluctuations quickly vanish, and the condensate establishes a global phase. The situation, however, is quite different in lower dimensions. The Mermin-Wagner-Hohenberg [60, 61] theorem states that condensation does not occur at finite temperatures in uniform one- (1d) and two-dimensional (2d) Bose gases. Studies of 1d traps [62, 63] with aspect ratios $\lambda = \omega_r/\omega_z \gg 1$ between the radial trap frequency ω_r and the longitudinal trap frequency ω_z showed suppressed density fluctuations at finite temperature $T \ll T_c$. However, along the length of the cloud phase fluctuations are present due to thermal excitations. The main contribution comes from axial excitations with energies $\epsilon_\nu < \hbar \omega_r$. Typically the range of the excitations' wavelengths are smaller than the length of the elongated BEC but larger than its diameter. In the vicinity of phase fluctuations we can no longer speak of a true BEC and we refer to such a system rather as a quasicondensate. The length scale on which phase fluctuations occur depends on the 1d density n_1 of the cloud and is defined by the phase coherence length

$$L_\phi = \frac{\hbar^2 n_1}{m k_B T}. \qquad (2.12)$$

In fact the phase coherence length increases with decreasing temperature T. Since a realistic confining potential is of finite size, a true BEC is observed as soon as L_ϕ becomes of the order of the length L of the BEC. We define the 1d transition temperature T_ϕ as the temperature where $L_\phi = L$ yielding

$$T_\phi = \frac{\hbar^2 n_1}{m k_B L}. \qquad (2.13)$$

At temperatures $T_\phi < T < T_c$ the coherence length is smaller than the condensate and causes the break up into ξ local domains of constant phase along the length of the cloud

$$\xi = \frac{T}{T_\phi} = \frac{L}{L_\phi}. \tag{2.14}$$

In terms of interferometry the locality of the phase immediately provokes the question whether a phase shift is actually measureable with such a system. Indeed we operate a magnetic trap in our experiment with an aspect ratio of around $\lambda = 71$ whereas Dettmer *et al.* [63] observed longitudinal phase fluctuations for much smaller ratios. Hence, the repeatability of a phase measurement is not trivial. An answer was given experimentally by G.-B. Jo *et al.* [23] in 2007. They demonstrated that a determination of the relative phase in a split BEC with an aspect ratio of ~ 200 is robust against phase fluctuations.

2.3 Theory of Matter Wave Interference

In this section we deal with the simple situation sketched in figure 2.1 where two BECs A and B that are initially well separated and independent overlap in free fall after being released from the trap at time $t = 0$. During their expansion the clouds establish their relative phase [64, 65] in the measurement of the resulting interference pattern. We define their relative phase as $\phi = \phi_a - \phi_b$, the difference of the individual phases $\phi_{a,b}$. We want to note that for simplification we refer to the relative phase often also only as phase within the text of this thesis.

2.3.1 Mean Field Picture

The simplest description of interference fringes [66] can be derived within the mean field picture, where $\Psi_{a,b}(\mathbf{r}, t) = \sqrt{n_{a,b}(\mathbf{r}, t)} e^{i\phi_{a,b}(\mathbf{r},t)}$ are the equilibrium wave functions of the two condensates. The initial separation of the condensates is $\mathbf{d}$, the condensate A being centered at $x_a = +d/2$, and the condensate B at $x_b = -d/2$ with $d = |\mathbf{d}|$. The initial order parameter of the system at time $t = 0$ is then written as a linear combination of the individual wavefunctions

$$\Psi(\mathbf{r}) = \Psi_a(\mathbf{r}) + \Psi_b(\mathbf{r}), \tag{2.15}$$

with the overlap of the wavefunctions

$$\int \Psi_a^*(\mathbf{r}) \Psi_b(\mathbf{r}) \, d^3r \cong 0. \tag{2.16}$$

If we neglect the interaction between the two condensates but account for the atom-atom interactions during free expansion, the total density $n = |\Psi|^2$ of the overlapping condensates exhibits modulations of the form

$$n(\mathbf{r}, t) = n_a(\mathbf{r}, t) + n_b(\mathbf{r}, t) + 2\sqrt{n_b(\mathbf{r}, t) n_b(\mathbf{r}, t)} \cos[\phi_a(\mathbf{r}, t) - \phi_b(\mathbf{r}, t)]. \quad (2.17)$$

By recalling that the phase of a condensate is related to the velocity field of a wavefunction via $\mathbf{v} = \frac{\hbar}{m}\nabla\phi$ and assuming that the velocity field of a condensate asymptotically approaches the classical velocity of the particles $\mathbf{v}_{a,b} = \mathbf{r}_{a,b}/t$ we find from integration the expression

$$\phi_{a,b}(\mathbf{r}, t) = \frac{1}{2}\frac{m}{\hbar t}(\mathbf{r} \pm \mathbf{d}) \cdot \mathbf{r} + \text{const.} \quad (2.18)$$

The difference between the two phases at the space point $\mathbf{r}$ then becomes

$$\phi_a(\mathbf{r}, t) - \phi_b(\mathbf{r}, t) = \frac{m}{\hbar t}\mathbf{d} \cdot \mathbf{r} + \phi. \quad (2.19)$$

One should notice that the integration constant has to be replaced by the relative phase ϕ which is an initial, constant offset inside the trap. Since ϕ has no physical significance with respect to a single cloud, the release of both clouds has the same properties and the relative phase is independent of the position in space. Hence equation (2.17) can be written as

$$n(\mathbf{r}, t, \phi) = n_a(\mathbf{r}, t) + n_b(\mathbf{r}, t) + 2\sqrt{n_b(\mathbf{r}, t) n_b(\mathbf{r}, t)} \cos\left(\frac{md}{\hbar t}x + \phi\right) \quad (2.20)$$

which shows that in a single experiment the interference pattern is characterized by straight line fringes orthogonal to the splitting axis with spacing

$$\Lambda = \frac{ht}{md} \quad (2.21)$$

between adjacent fringes. Since the absolute position of the fringes depends on the initial relative phase ϕ, the relative phase can be determined from the density distribution of two overlapping condensates. This is the basic principle of a phase measurement used for evaluating images from the interferometer. How this works in practice is presented in chapter 6.

2.3.2 Two Mode Theory

The mean field approximation predicts the occurrance of interference and describes well the fringe spacing Λ. However, it makes no comment about how the density measurements establish the relative phase ϕ. In order to address this we introduce the two mode approximation which has been extensively discussed in the literature and is also used to describe the relation between phase and relative atom number in double well systems [67]. In the two mode expansion the field operators are expanded in terms of two mode functions $\psi_i(\mathbf{r}, t)$ with i = a,b. The model works if higher excitation modes can be neglected and the restriction to two modes is adequate for most aspects of the problems with weak atom-atom interactions [68].

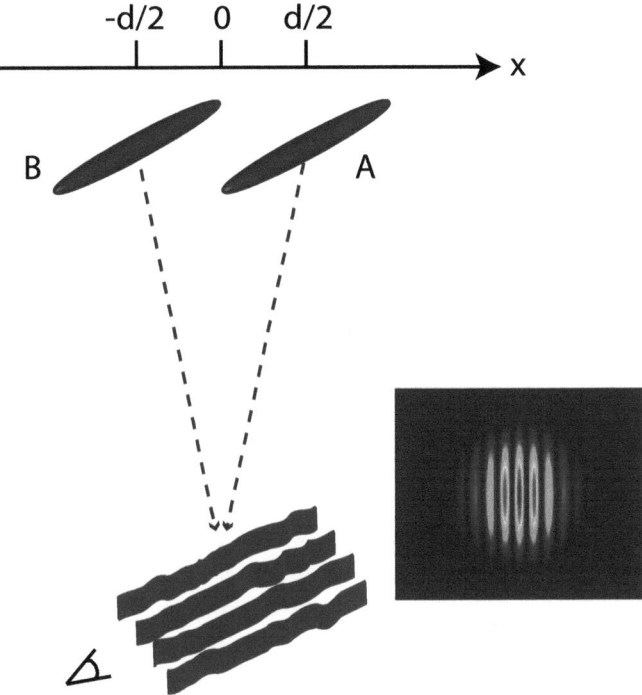

Figure 2.1: Schematic of two interfering BECs. The elongated BECs have an initial separation d and overlap in during time-of-flight. An observer looking along the length of the cloud can see the characteristic interference pattern. Ideally a fringe visibility of 100% is expected. In experiments, however, phase fluctuations along the length of elongated BECs lead to a reduced contrast.

The two mode theory defines the annihilation $\hat{a}_i(t)$ and creation operators $\hat{a}_i^\dagger(t)$ such that at a point in time t

$$\hat{a}_{a,b}^\dagger(t) = \int \psi_{a,b}(\mathbf{r},t)\hat{\Psi}^\dagger(\mathbf{r})\,d^3r. \tag{2.22}$$

The time-dependent definition ensures that the theory is also capable of describing the physics after the BECs are released from the trap. We have then just to put in the correct single particle wavefunction at time t. The bosonic field operators $\hat{\Psi}(\mathbf{r})$ satisfy the standard commutation relations

$$\left[\hat{\Psi}(\mathbf{r}'),\hat{\Psi}(\mathbf{r})\right] = \left[\hat{\Psi}^\dagger(\mathbf{r}'),\hat{\Psi}^\dagger(\mathbf{r})\right] = 0,$$
$$\left[\hat{\Psi}(\mathbf{r}'),\hat{\Psi}^\dagger(\mathbf{r})\right] = \delta(\mathbf{r}'-\mathbf{r}). \tag{2.23}$$

In analogy to equation (2.8) we can write the field operator in terms of a relative density $\hat{n}$ and relative phase operator $\hat{\phi}$

$$\hat{\Psi}(\mathbf{r}) = \sqrt{\hat{n}(\mathbf{r})}e^{i\hat{\phi}(\mathbf{r})}. \tag{2.24}$$

Reproduction of the commutation relations (2.23) implies

$$[\hat{n}(\mathbf{r}), \hat{\phi}(\mathbf{r}')] = i\delta(\mathbf{r} - \mathbf{r}') \qquad (2.25)$$

which means that the density and phase operator are conjugate variables. From the fact that the two operators do not commute follows immediately an uncertainty relation. If the relative atom number between two BECs is well-defined, then the relative phase appears to be completely random.

Number States

In the two mode picture we can define the so-called relative number states

$$|k, N\rangle = \frac{\left(\hat{a}_a^\dagger\right)^{\left(\frac{N}{2}-k\right)} \left(\hat{a}_b^\dagger\right)^{\left(\frac{N}{2}+k\right)}}{\left[\left(\frac{N}{2}-k\right)!\right]^{\frac{1}{2}} \left[\left(\frac{N}{2}+k\right)!\right]^{\frac{1}{2}}} |0\rangle \qquad (2.26)$$

where N is the total number of atoms in the system and k can have the values $-N/2$, $-N/2 + 1$,..., $+N/2$. The atoms occupy two independent modes which we define as a fragmented state. In such a state, also called Fock state, the number of atoms in each mode is well-defined and it therefore immediately follows from (2.25) that the absolute phase is not well-defined.

We want to examine the interference properties of such a state when released from the trap. After a long enough time-of-flight in the regime that the two clouds overlap we can write the single particle wavefunctions as a Gaussian function

$$\psi_{a,b} = u_{a,b}(\mathbf{r}, t) e^{i\mathbf{Q}_{a,b} \cdot \mathbf{r}} \qquad (2.27)$$

where $\mathbf{Q}_{a,b} = m(\mathbf{r} \pm \mathbf{d}/2)/\hbar t$ are the wavevectors and $u_{a,b}$ are slowly varying real functions describing the overall density. The strong overlap of the mode functions yields the following normalization

$$\int u_{a,b}(\mathbf{r}, t)^2 \, d^3r = 1, \qquad (2.28)$$

$$\int u_a(\mathbf{r}, t) u_b(\mathbf{r}, t) \, d^3r \approx 1. \qquad (2.29)$$

Asking for the amplitude of density oscillations at wavevector $\mathbf{Q} = \mathbf{Q}_a - \mathbf{Q}_b$ the associated operator in second quantization is given by

$$\hat{\rho}_\mathbf{Q} = \int \hat{\Psi}^\dagger(\mathbf{r}) \hat{\Psi}(\mathbf{r}) e^{i\mathbf{Q} \cdot \mathbf{r}} d^3r \qquad (2.30)$$

which is the Fourier transform of the number density operator at the corresponding wave vector. The expectation value of the operator $\hat{\rho}_\mathbf{Q}$ can be calculated explicitly for the number state defined in equation (2.26)

$$\langle k, N | \hat{\rho}_\mathbf{Q} | k, N \rangle = \left\langle k, N \left| \int \hat{\Psi}^\dagger(\mathbf{r}) \hat{\Psi}(\mathbf{r}) e^{i\mathbf{Q} \cdot \mathbf{r}} d^3r \right| k, N \right\rangle. \qquad (2.31)$$

We evaluate the integral by assuming that the extension of the clouds $u_{a,b}$ varies on a scale longer than the fringe wavelength $1/|\mathbf{Q}|$. Evaluation of the Fourier integral annihilates then all parts with a wavevector different from $\mathbf{Q}$ and together with the normalization in equation (2.28) and (2.29) we find

$$\langle k, N | \hat{\rho}_Q | k, N \rangle = \langle k, N | \hat{a}_a^\dagger \hat{a}_b | k, N \rangle = 0. \tag{2.32}$$

This result seems to suggest that there are no density modulations for number states. A single shot experiment, however, clearly exhibits an interference pattern consistent with equation (2.20). This arises because the expectation value in quantum mechanics has to be interpreted as the statistical average over many experiments. For a Fock state the relative phase is completely random [69] and the repeated and averaged experiment recovers the result of equation (2.32). A Fock state would therefore constitute an incoherent input state making it completely unsuitable for interferometry.

Phase States

In order to describe a many-body state with a well-defined phase we need to introduce the binomial, or so-called phase states,

$$|\phi, N\rangle = \frac{1}{\sqrt{N!2^N}} \left(\hat{a}_a^\dagger + e^{-i\phi} \hat{a}_b^\dagger \right)^N |0\rangle. \tag{2.33}$$

All atoms occupy the same single particle state $(\psi_a + e^{-i\phi}\psi_b)/\sqrt{2}$ which is a superposition of the two mode functions. This is called an unfragmented BEC. Although the total number of atoms is fixed, the relative atom number k is completely uncertain. The relative phase ϕ is constant and the calculation of the expectation value of the density operator $\hat{\rho}_Q$ yields

$$\langle \phi, N | \hat{\rho}_Q | \phi, N \rangle = \frac{N}{2} e^{-i\phi}. \tag{2.34}$$

This result suggests that the interference pattern develops with a fixed phase ϕ. The crucial point is that the average of repeated experiments delivers the same result as a single shot. In fact for a pure binomial state $\langle \hat{\phi} \rangle = \phi$ and $\langle \Delta \phi^2(t) \rangle = \langle \phi, N | \hat{\phi}^2 | \phi, N \rangle - (\langle \phi, N | \hat{\phi} | \phi, N \rangle)^2 = 0$.

Phase states represent a basis of the many-particle Hilbert space, thus we can write any state as a superposition of phase states. For example we can write the Fock state as

$$|k, N\rangle = \int_{-\pi}^{+\pi} \chi(\phi) |\phi, N\rangle \, d\phi \tag{2.35}$$

where $\chi(\phi) = \frac{1}{2\pi}$ is the distribution of the phase states and is called the phase amplitude. In this picture one can interpret the phase measurement as the projection onto a random phase state with a probability given by $\chi(\phi)$. The ensemble average would, however, still be given by (2.32).

Ensemble Average

In this section we discussed what we can expect from repeated experiments with the two extremes of number states and binomial states. Let us now consider which state we end up with when splitting the BEC in a double well potential. It is possible that the distribution of phase states $\chi(\phi)$ is completely different, i.e. we can assume a Gaussian time-dependent phase amplitude [70]

$$\chi(\phi, t) = \frac{1}{(2\pi \langle \Delta\phi^2(t)\rangle)^{1/4}} \exp\left(-\frac{\phi^2}{4\langle \Delta\phi^2(t)\rangle} + \frac{i\delta(t)}{2}\phi^2\right), \qquad (2.36)$$

where $\delta(t)$ is a function in time. We are now able to calculate the averaged ensemble density for an arbitrary phase spread $\langle \Delta\phi^2(t)\rangle$ by

$$n(\mathbf{r},t) = \int_{-\pi}^{+\pi} n(\mathbf{r},t,\phi) |\chi(\phi,t)|^2 \, d\phi = \frac{1}{2}\left[n_a(\mathbf{r},t) + n_b(\mathbf{r},t) + n_{\text{int}}(\mathbf{r},t)\right] \qquad (2.37)$$

with the interference term

$$n_{\text{int}}(\mathbf{r},t) = 2\sqrt{n_a(\mathbf{r},t)n_b(\mathbf{r},t)} \cos\left(\frac{m}{\hbar t}\mathbf{d}\cdot\mathbf{r} + \phi\right) e^{-\langle \Delta\phi^2(t)\rangle/2}. \qquad (2.38)$$

In the averaged ensemble the fringe contrast is a measure of the phase distribution. The highest contrast we find again for the binomial state for which $\langle \Delta\phi^2(t)\rangle = 0$. The ensemble average is hence a tool to reconstruct the quantum state of the double well potential and tells us how close we are to the binomial or number state after the preparation of the atoms.

2.3.3 Dephasing of a Coherent State

The simplest model [66] of our experiment is to assume that each atom makes a choice of occupying the left or right well. This splits the initial BEC into a coherent state,

$$|\phi, N\rangle = e^{-iN\phi/2} \sum_k c(k) \, e^{ik\phi} \, |k, N\rangle. \qquad (2.39)$$

The probability distribution $|c(k)|^2$ is bionomial. The coefficients are

$$c(k) = \frac{\sqrt{N!}}{\sqrt{2^N (N/2+k)! (N/2-k)!}}. \qquad (2.40)$$

Each number state $|k, N\rangle$ differs in energy due to atom-atom interactions which depends on the number of atoms. Therefore each state evolves with a phase factor $e^{-iE(k)t/\hbar}$ and by expansion of the number state energy $E(k)$ around $k = 0$ we find [71]

$$|\Psi(t)\rangle = e^{-iE_0 t/\hbar} \sum_k c(k) \, e^{-iE_c k^2 t/\hbar} \, |k, N\rangle, \qquad (2.41)$$

where E_0 is the ground state energy of the double well potential and $E_c = \frac{\partial^2 E(k)}{\partial k^2}|_{k=0}$ the on-site energy. The quantum state is a sum of number states each evolving with its own phase. Since each phase evolves differently, the width of the phase distribution increases in time. For an ideal gas $E_c = 0$ which means the phase spreading arises from atom-atom interactions. The phase spread $\langle \Delta\phi(t) \rangle$ increases with time from the initial phase distribution $\sigma_{\phi,0}$ at the end of the splitting process according to

$$\left\langle \Delta\phi^2(t) \right\rangle = \sigma_{\phi,0}^2 + \sigma_k^2 \left(\frac{E_c}{\hbar} t \right)^2. \tag{2.42}$$

The phase spreading is proportional to the fluctuations in relative atom number σ_k^2 which remains fixed as there is no tunneling. The uncertainty in relative atom number is accompanied by an uncertainty in the relative chemical potential $\sigma_{\Delta\mu} = \frac{\partial\mu(k)}{\partial k}|_{k=0}\sigma_k$ because the chemical potential μ on one site is linked to the energy of a number state via $E(k) = \int \mu(k)\, dk$.

The dephasing time t_d is defined by the relation $\left\langle \Delta\phi^2(t_d) \right\rangle \cong 1$ rad [64, 66]. A state with an initially well-defined phase $\sigma_{\phi,0} \ll 1$ has therefore a dephasing time of

$$t_d = \frac{\hbar}{\sigma_{\Delta\mu}}. \tag{2.43}$$

Assuming Poissonian number fluctuations $\sigma_k = \sqrt{N}$ the Thomas-Fermi model [72] estimates an uncertainty of

$$\sigma_{\Delta\mu} = \hbar \left(\frac{72}{125} \frac{m}{\hbar} \right)^{1/5} \frac{\omega^{6/5} a_s^{2/5}}{N^{1/10}} \tag{2.44}$$

where ω is the geometric mean frequency of the trap and a_s the s-wave scattering length. Hence, the dephasing time depends weakly on the atom number with $t_d \propto N^{1/10}$.

2.3.4 Contrast in Elongated Systems

So far we have only considered the interference of 3-dimensional (3d) systems with a global phase. Theory predicts for interfering clouds in a coherent state a fringe contrast of 100%. Low frequency fluctuations of longitudinal BECs, however, lead to phase fluctuations along the length of the cloud. Splitting a BEC creates two identical clouds with zero relative phase and despite the locality of the phase, interference fringes are aligned and clearly visible along the length of the clouds [73]. Phase spreading, however, increases the uncertainty of a local phase measurement and the position of the fringes varies from domain to domain. Assuming the clouds are aligned along the z axis, and that we observe the system along the same axis, imaging then represents an integration over the imaged area with a length L. Hence the fringe contrast is reduced due to fluctuations in the phase difference. Therefore phase diffusion with time is relevant for the fringe contrast in elongated systems.

The fringe amplitude is again given by the operator $\hat{\rho}_Q$ of equation (2.30) whereas for a 1d system the reduced dimensionality requires only the integration along the z direction, and the operators $\hat{a}_{a,b}(z)$ and $\hat{a}^\dagger_{a,b}(z)$ are z dependent. In the limit of a large atom number N within the integration length L the expectation value of $\hat{\rho}_Q$ is given by the expression [74]

$$\left\langle |\hat{\rho}_Q(L)|^2 \right\rangle = \int_0^L \int_0^L \left\langle \hat{a}_a^\dagger(z)\hat{a}_b^\dagger(z')\hat{a}_a(z')\hat{a}_b(z) \right\rangle dz\,dz'. \tag{2.45}$$

Since the clouds are split into a symmetric double well potential, the clouds are identical and have the same atom density with equal interaction coupling. Equation (2.45) simplifies then to

$$\left\langle |\hat{\rho}_Q(L)|^2 \right\rangle = \int_0^L \int_0^L \left\langle \hat{a}^\dagger(z)\hat{a}(z') \right\rangle^2 dz\,dz' \tag{2.46}$$

with $\hat{a}(z) = \hat{a}_{a,b}(z)$ and $\hat{a}^\dagger(z) = \hat{a}_{a,b}^\dagger(z)$. This equation allows us to have a qualitative discussion about the influence of phase fluctuations on the contrast. For small coherence length $L_\phi \ll L$ the expectation value $\left\langle \hat{a}^\dagger(z)\hat{a}(0) \right\rangle$ decays fast with distance z. Assuming an exponential decay yields the scaling $|\hat{\rho}_Q(L)| \propto \sqrt{L_\phi L}$ and the smaller the coherence length the smaller the fringe contrast. In the opposite regime that the coherence length is on the order of the integration length $L_\phi \sim L$ the correlation function $\left\langle \hat{a}^\dagger(z)\hat{a}(0) \right\rangle$ must be constant and hence the fringe amplitude is $|\hat{\rho}_Q(L)| \propto L$ depending only on the integration length.

Chapter 3

Experimental Requirements

3.1 Introduction

Atom chips have greatly simplified many atomic physics experiments. Despite this the complexity of such experiments remains high. Although there have been attempts to build more compact BEC experiments [75], our experiment still requires an entire optical table. The main obstacle is to combine the different technologies required for condensation, i.e. lasers, vacuum, electronics etc., and to coordinate a multitude of operations with μs precision.

In this chapter we present the experimental setup and describe how we achieve a BEC on an atom chip. The subject is also covered in earlier theses on our apparatus [76, 77, 78] where the parts of the hardware and the experimental cycle are discussed in more detail. The purpose of this chapter is to give the reader an idea of the basic requirements necessary to achieve condensation. We present the infrastructure and the individual components of the system in section 3.2. Especially, the implementation of the rf generation is of importance for the later chapters in this thesis. Afterwards we briefly discuss the experimental cycle and the control software in section 3.3. The different steps and the sequence to achieve condensation are explained.

3.2 Infrastructure

3.2.1 Vacuum System

Ultra-high vacuum (UHV) is necessary to achieve BEC transition and long magnetic trap lifetimes. A schematic of the vacuum setup is presented in figure 3.1. It consists of two parts, the low velocity intense source (LVIS) chamber and the main chamber, separated by a small aperture and a gate valve. Inside the vacuum chambers only clean and UHV-compatible components are permitted, since contaminated parts would outgas preventing UHV. In order to remove water from the parts inside the chamber we baked the system at temperatures chosen low enough not to destroy the chip assembly. Following a bake out at 150°C for one week we eventually obtained a pressure of less than 10^{-11} Torr in the main chamber after two month of pumping. Especially in the science chamber where the atom chip is located an extremely low pressure is essential to achieve BEC.

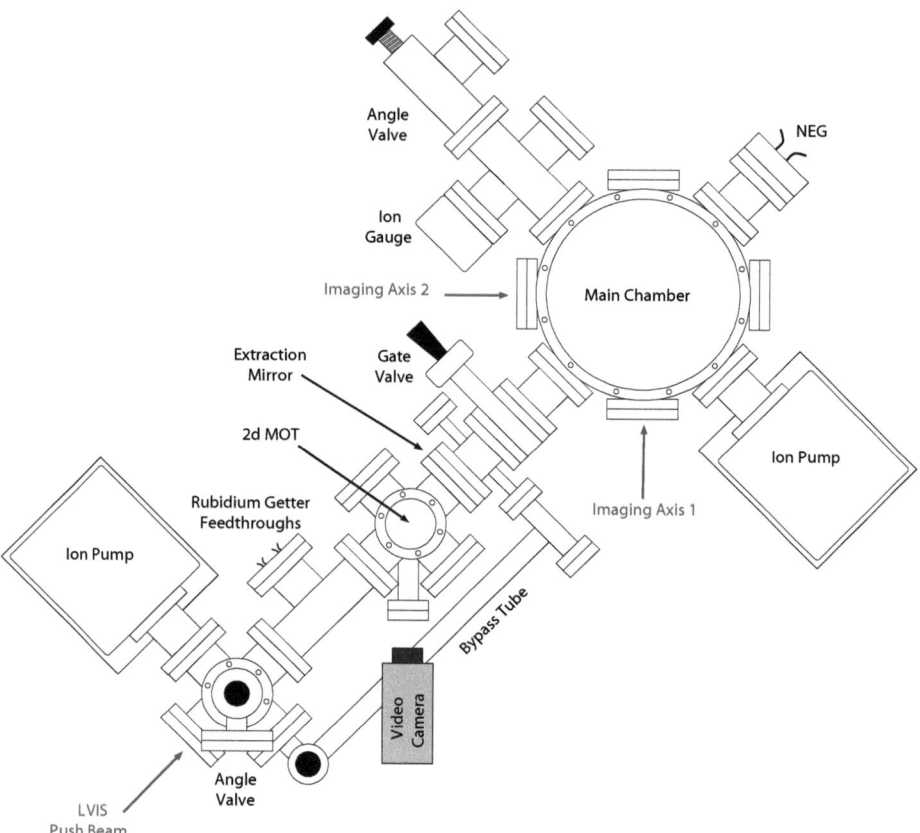

Figure 3.1: The vacuum system contains the experimental platform. The two main parts, the science chamber and the LVIS chamber are connected via a 1 mm pinhole in the extraction mirror. The pressure in each chamber is maintained by an ion pump. Viewports allow optical access for the MOT, optical pumping and imaging beams.

LVIS Chamber

The LVIS chamber serves as a 2d MOT providing slow atoms for the experiments on the atom chip. This LVIS allows fast loading of the MOT while maintaining low vacuum pressure in the science chamber. The LVIS design is based on the system described in C. Sinclair's thesis [79]. The LVIS vacuum chamber consists of two $2\frac{3}{4}$ in conflat six-way crosses connected by a T-piece. The four viewports of the cross close to the science chamber give access to four laser beams in order to create a 2d MOT. A push beam enters the viewport at the second cross perpendicular to the other beam paths. The push beam is retro-reflected on a mirror with a 1 mm aperture through which the atoms are released into the science chamber. The aperture connects both parts of the vacuum system with a conductance of $0.02\,\mathrm{ls}^{-1}$. A Varian VacIon Plus 20 ion pump with a pumping rate of $20\,\mathrm{ls}^{-1}$ maintains the vacuum pressure in the LVIS system. A soft iron shell around the ion pump shields its large magnetic fields from the experiment. Two rubidium dispensers are mounted in the flange of the T-piece. We keep the dispenser in use warm by permanently running a current of 2 A through them. At a current of around 4.5 A rubidium vapor is released into the chamber allowing loading of the LVIS MOT.

Main Chamber

The main vacuum chamber has a spherical-octagon shape and is made from 304 grade stainless steel by Kimball Physics Inc. Four of the eight CF40 conflat ports around the perimeter are equipped with viewports. Together with the large CF160 viewport at the bottom of the chamber these provide sufficient optical access for the MOT, imaging, optical pumping beams and visual inspection. The top of the chamber is sealed with a CF160 flange on which the chip mounting is assembled. The chip on its mount hangs upside down in the science chamber. The top flange also contains three multi pin feedthroughs to make electrical connections to the outside world.

To maintain UHV we use a $20\,\mathrm{ls}^{-1}$ Varian VacIon Plus 20 ion pump and a non-evaporative getter (NEG) made of an Al-Zr alloy. The remaining CF40 port is equipped with a Lesker VZCR40R angle valve providing the possibility of attaching a turbo pump to the system. We record the pressure with an ion gauge mounted on a T-piece between angle valve and science chamber.

3.2.2 The Atom Chip

The atom chip shown in figure 3.2 consists of a $3\,\mu\mathrm{m}$-thick gold layer evaporated onto a p-doped silicon wafer cut along the [100] axis. Silicon has an advantage over other substrates because of its high thermal conductivity. The substrate has a resistivity of 17-30 Ωcm and must therefore be electrically isolated from the chip wires to avoid cross connections. This is achieved with a 100 nm layer of SiO_2 between the gold and the silicon. The oxide is kept thin to avoid reduction reducing the thermal conduction. The wire pattern on the surface is defined by ion beam milling. Special attention must be paid to wire defects as the surface quality and edge roughness can lead to inhomogenous current densities [77] which introduce anomalous magnetic fields along the length of the trap. The fabrication process is described in detail in [78] and our fabrication paper [80] where the characterization of the chip wires is also discussed.

Wire Pattern Design

The layout of the chip wires is relatively simple and is presented in figure 3.2. There are four parallel Z-wires, labelled Z1 to Z4, in the centre of the chip that are used to produce the necessary DC and rf fields for trapping and manipulating BECs near the surface of the chip. The wires in the outer pair are $100\,\mu\mathrm{m}$ wide and have a centre to centre separation of $300\,\mu\mathrm{m}$. The inner wires are $50\,\mu\mathrm{m}$ wide

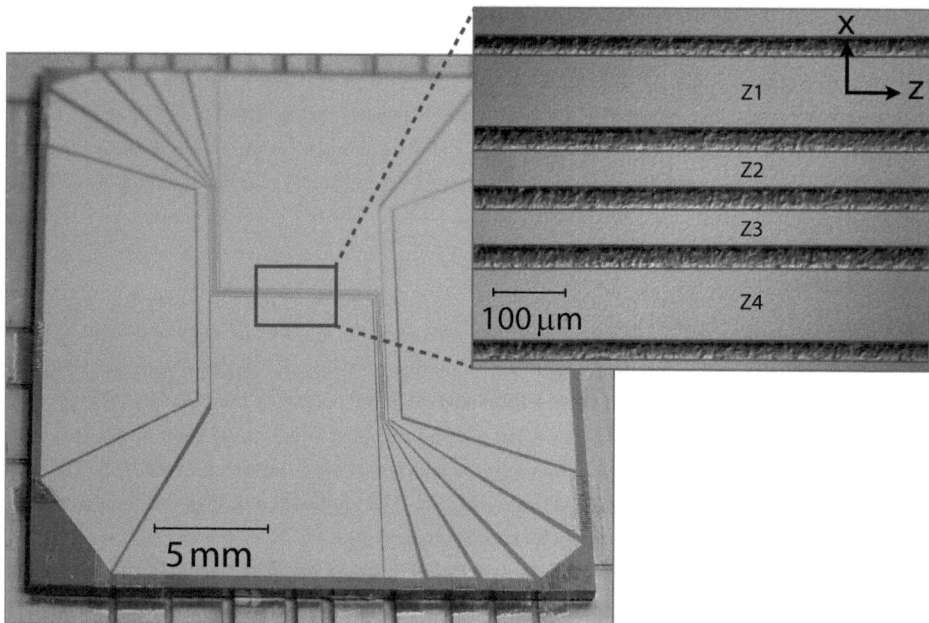

Figure 3.2: The atom chip used in our experiment. The four Z-shaped wires in the centre are clearly visible. Two of them have a width of 50 μm, the other two are 100 μm wide. The inset shows a magnified image of the central region. The Z-wires carry both DC and rf currents to create the magnetic trap and control the splitting into a double well potential. The additional wires parallel to the ends of the Z-wires can carry large currents to provide additional axial confinement of the trap. The highly reflecting gold pads act as a mirror for the mirror MOT in the early stage of atom trapping.

and have a separation of 85 μm. The central section of the wires along the z axis, above which the BEC is trapped, is 7 mm long. Two more wires are patterned onto the chip parallel to the ends of the Z-wires along the x axis. These so-called endwires can provide an additional field to adjust the frequency of the magnetic trap along the z axis and play a key role for loading the magnetic trap. The large gold pads on the chip surface are highly reflective and serve as a mirror for the reflection MOT in the initial, experimental stage.

In a previous series of experiments two of the chip wires were accidentally damaged by running excessive currents for too long, leaving only one of the inner and one of the outer Z-wires available for use. This also reduced the cross resistance between the two wires to around 200 Ω. All the experiments presented here are therefore realized with two unequal wires, labelled Z3 and Z4 respectively. The centre-to-centre separation for these two is 107.5 μm. The resistance of Z3 is 4.6 Ω while Z4 has a resistance of 2.4 Ω due to the larger cross sectional area.

Chip Mounting

The position of the atom chip inside the vacuum chamber is defined by the chip mount which must be mechanically stable and should not drift during the lifetime of the experiment. It also provides the electrical connections for the chip and acts as a heat sink.

We mount the atom chip directly onto a Shapal-M base plate using two copper clamps. Shapal-M is a glass ceramic with a high thermal conductivity ($90\,\mathrm{Wm^{-1}K^{-1}}$) which is easy to machine. Embedded into the Shapal-M plate is a copper structure which has direct contact with the back of the atom chip. The structure has the shape of an H with a 2 mm square cross section and a central cross piece of 2 mm length and is glued in place with UHV compatible epoxy (Bylapox). The H is centred on the chip surface and serves two purposes. First, it dissipates heat away from the wires and second, we can run currents through it to provide additional magnetic fields, both static and rf. In our experiments we run small rf currents through the copper-H to implement evaporative cooling or to perform rf spectroscopy on the magnetically trapped atoms.

To mount the whole structure inside the chamber, as shown in figure 3.3 (left), the Shapal-M is glued to a 7 mm thick copper block which is in turn connected to the vacuum flange by three copper legs with a length of 21.5 mm. The copper legs are directly screwed onto the copper block and the flange permitting high thermal dissipation to the vacuum chamber which serves as a heat sink. Together with the legs we also assembled the holder for the two MOT coils shown in figure 3.3 (right). The MOT coils provide a magnetic quadrupole field centred a few mm above the chip surface. The holder is a bent metal sheet that holds one coil former below, and one above, the atom chip.

Since it is difficult to solder to a gold surface, the electrical connection to the chip wires is made by clips. The broad pads at the chip edge are especially designed for this purpose. The clips are directly plugged onto the chip and not further held by glue. They are supported by two Macor blocks screwed to each other. These clips are commercially made from a copper-tin alloy $CuSn_6$ by Batten&Allen. In order to connect to the outside world copper wires are attached to vacuum feedthroughs with barrel connectors. On the chip side the wires are soldered directly to the clips using UHV-compatible, fluxless solder from Allectra.

3.2.3 Laser System

In order to cool and prepare the atoms for magnetic trapping we need various laser beams locked to the rubidium D2 transition. For this purpose we have four different lasers providing the light for the MOT, imaging and optical pumping. A detailed description of the laser setup and the locking techniques can be found in M. P. A. Jones' thesis [76]. A schematic diagram is given in figure 3.4.

The Lasers

The trapping and imaging light comes from a Coherent MBR-110 Ti:Sapphire laser with an output power of up to 1 W linearly polarized light at 780 nm wavelength. The MBR-110 consists of a ring cavity with a Ti:Sapphire crystal pumped by laser light at 532 nm. An 8 W Verdi laser from Coherent

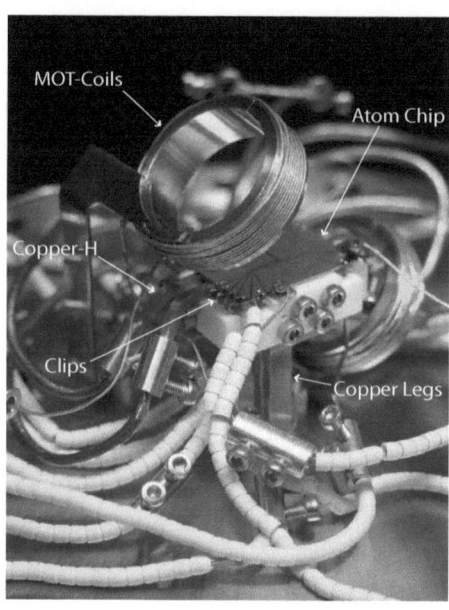

Figure 3.3: Left: Schematic of the chip mount. Right: Picture of the chip mount assembled to the top flange just before putting it into the vacuum chamber.

produces the necessary pump beam. The MBR-110 possesses a reference cavity for locking. The error signal for the laser lock is derived from a beat signal with the reference laser locked to the $F_g = 2 \rightarrow F_e = 3$ transition. This results in a stable output frequency of the MBR-110 which is 100 MHz blue-detuned to the same transition

The reference laser and the repump laser are home-made external cavity diode lasers in Littrow configuration [81]. Both have a mode hop free scan range of around 3 GHz and an output power of 30 mW at a wavelength of 780 nm. Beside the beat signal the reference laser also provides light for the optical pumping beam. The repump laser is needed to close the loss channel in the laser cooling cycle and is thus locked to the $F_g = 1 \rightarrow F_e = 2$ transition.

MOT Beams

For the MOT we red-detune laser light from the MBR-110 by 13 MHz with an acousto-optical modulator (AOM) and overlap it with the repump light in a polarizing beam cube. An additional shutter after the AOM ensures the blocking of the beam. The laser light is telescoped after the cube and splits into the beam paths for the diagonal MOT beams, the horizontal MOT beams and the LVIS MOT beams. The diagonal MOT beams enter the chamber at the bottom through the big CF160 viewport and are retro-reflected under an angle of 45° on the atom chip surface. The horizontal MOT beams enter through two opposite viewports travelling parallel to the chip surface in z direction. A $\lambda/4$ wave-

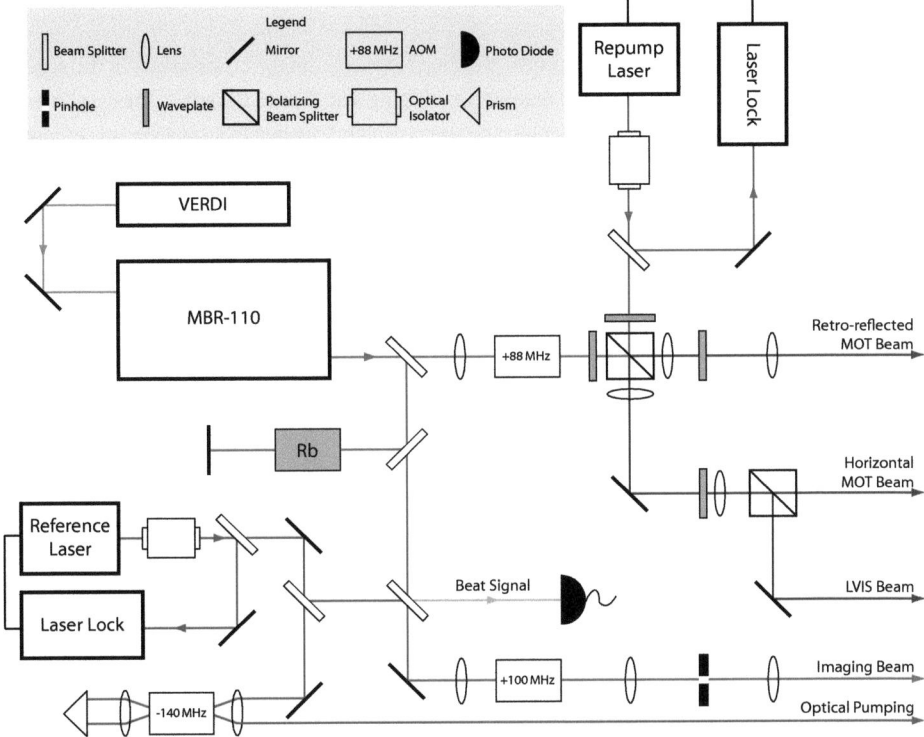

Figure 3.4: The laser setup includes four different lasers. The MBR-110 is pumped by a 8 W Verdi laser and generates the trapping light for the MOT and the imaging beams. A beat lock with the reference diode laser, also used for optical pumping, assures frequency stability. The exact detuning of each beam is set by an AOM in the beam path. To close the cooling cycle another diode laser provides repump light which is overlapped with the trapping beams in a beam cube.

plate in the beam path of each MOT beam shifts the linearly polarized light into the required circular polarisation.

Optical Pump Beam

The optical pumping beam is derived from the reference laser, double passed through a 140 MHz AOM. The switching of the light is controlled by both the AOM and an additional shutter. Its frequency is 13 MHz red-detuned from the $F_g = 2 \rightarrow F_e = 2$ transition and is overlapped with one of the diagonal MOT beams to pass it onto the atoms.

3.2.4 Imaging System

In order to get information out of the system we have to take an image of the atoms in the state of interest. Since our imaging technique is destructive, imaging also defines the end of the experimental cycle. A complete picture of an experimental outcome is usually gathered from repeated experiments. Therefore a stable and reliable setup of the imaging system is essential.

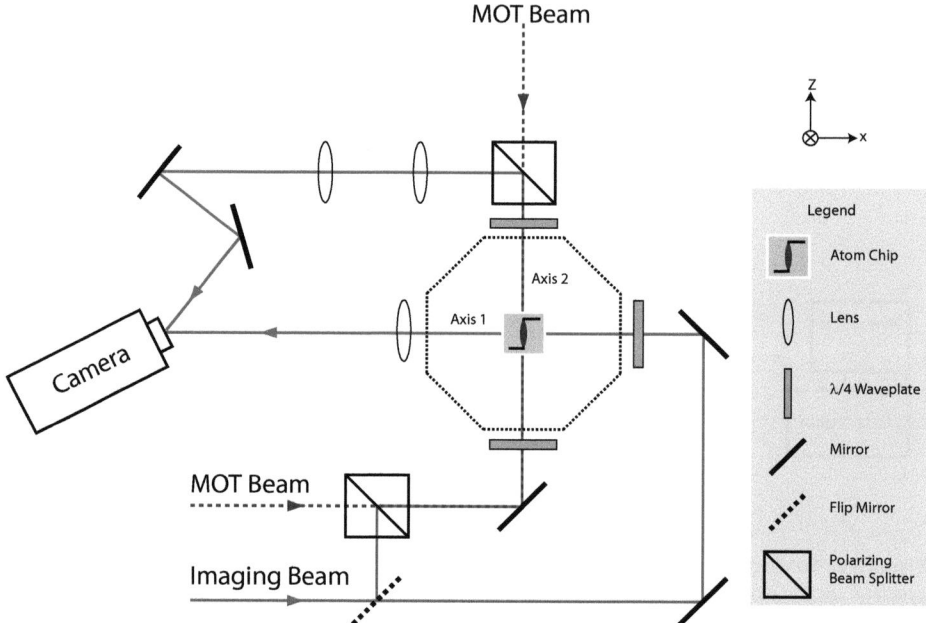

Figure 3.5: The imaging system provides two axes to observe the BEC on the chip. A flip mirror allows the selection between axis 1 and 2. Imaging axis 2 is orientated along the length of the trapped BEC and needs to be overlapped with the horizontal MOT beams. We use polarizing beam splitters for both to combine and separate the two laser beams. The imaging beams on both axes hit the camera under an angle of 26° to avoid etalon effects.

Imaging Beam

We image the atoms inside the vacuum chamber using absorption imaging [82] with a laser beam resonant on the $F_g = 2 \rightarrow F_e = 3$ transition of the rubidium D2 line. The pulse duration of $30\,\mu$s and the frequency of the beam is set by the AOM in the beam path. After the AOM the imaging beam is linearly polarized and passes through a spatial filter with a $40\,\mu$m pinhole. We adjust the intensity with a $\lambda/2$ plate and a polarizing beam splitter. A flip mirror allows the choice of two different imaging axes, marked as 1 and 2 in figure 3.1.

Imaging Axes

On imaging axis 1 we image perpendicular to the Z-wires along the x axis (see figure 3.1). A $\lambda/4$ waveplate in front of the vacuum chamber sets the polarisation to σ^+ with respect to the imaging field along the x direction. After passing through the chamber the imaging beam is focussed onto the camera chip by a single spherical lens in a 2f-2f configuration. The lens consists of a pair of achromatic doublets placed back to back with a focal length of 80.5 mm from Comar. The magnification on axis 1 is $M = 1.2$. We can capture fluorescence images of both the MOT and UMOT using this axis as well as absorption images of the trapped and dropped BEC.

Imaging on axis 2 coincides with the z axis of the chip. This imaging beam has to be overlapped with the horizontal MOT beams using a polarizing beam splitter. The MOT setup already requires a $\lambda/4$ waveplate in front and behind the vacuum chamber in order to achieve circularly polarized light. The first waveplate where the imaging beam enters the chamber changes its polarization to σ^+ with respect to a magnetic field along the z axis. The second $\lambda/4$ waveplate sets the initial polarisation and we are able to separate the imaging beam from the MOT beams with another polarizing beam splitter. The image is focussed onto the camera by two spherical lenses, the Melles-Griot 01 LAO 625 with a nominal focal length of $f = 200$ mm and the Melles-Griot 01 LAO 688 with a focal length of $f = 400$ mm. Imaging axis 2 has a magnification of $M = 1.67$. The interference patterns of two overlapping BECs can be observed on this axis.

Camera

We use a PentaMAX camera from Princeton Instruments to image our atoms. The heart of the camera is a CCD chip from Kodak which has 1317×1035 pixels each with an area of $6.8\,\mu\text{m} \times 6.8\,\mu\text{m}$ providing 0.026 counts per photon. The camera saturates at 4095 counts per pixel. For a reduced background signal the CCD chip is permanently cooled to $-30°$C. All camera parameters are set in the WinView Software from Princeton Instruments. Our control software initializes the camera at the beginning of each experimental cycle. A TTL trigger times the exposure of the CCD. Typically a received trigger opens the mechanical camera shutter for 110 ms. WinView then reads in the raw images and saves them into a file on the hard drive for further processing. To avoid etalon effects on the entrance window the camera is tilted by an angle of $26°$ in the horizonatal plane with respect to the imaging axis.

3.2.5 External Magnetic Fields

In order to create and control the MOT and the magnetic trap several external magnetic fields are required. With the exception of the quadrupole field generated by the LVIS MOT and the MOT coils, which are in anti-Helmholtz configuration, all other external magnetic fields are homogenous.

The two LVIS MOT coils have 100 turns of kapton-insulated copper wire each and are mounted onto the two vertical viewports of the LVIS chamber with a mean separation of 6.5 cm. For a working LVIS we typically run 2 A through each of the coils resulting in a field gradient of $15\,\text{Gcm}^{-1}$. The same current in the MOT coils creates a magnetic field of around $11\,\text{Gcm}^{-1}$. The MOT coils are mounted inside the chamber along an axis at 45° to the chip surface. The MOT coils are wound onto a stainless steel coil former with a 12.5 mm inner radius and a 45 mm centre to centre separation. A cut through each former prevents the flow of induced current when the coils are switched on or off.

Apart from the chip wire fields two external magnetic fields are important to create a magnetic trap, the X-bias field pointing along the x axis and a quantization field along the z axis. The former determines the distance of the trap from the chip surface while the latter defines the trap bottom and prevents spin flips. Both fields are produced by coils mounted outside the science chamber.

Additional coils provide shim fields along the x and y direction. These fields are primarily needed to optimise the loading of the trap and to define a quantization axis for imaging.

3.2.6 Current Control

The stability of the magnetic trap depends strongly on the noise level in the currents. To get stable currents independent of the load, we use low-noise power supplies and custom-made current controllers.

Power Supplies

For all currents running through the coils and chip wires we use commercial power supplies from TTi Ltd. and ISO-Tech. The current for the chip wires is delivered from a TTi EX354R with a ripple voltage < 2 mV rms.

Current Controller

The design of our home-made current controllers is presented in figure 3.6. The current through the chip wires is controlled by high current op-amps of the type OPA548 and OPA549. The output current is set by an analogue control voltage from the experiment computer. The integrated circuit LF412 provides two additional op-amps in front of the OPA548(9). The whole circuit has a common ground but the LF412 has a separate power supply. The first op-amp compares the control voltage with the voltage drop over a 1 Ω sense resistor R4 connected in series with the chip wires. A trimmer circuit takes a voltage drop over a potentiometer and adds it to the control voltage as an offset. This enables us to compensate the small offset current which arises due to the imperfect behaviour of op-amps [83]. The second op-amp operates as an integrator with a variable gain to stabilise the circuit. The integrator gain dominates the turn off behaviour of the chip wire current. Indeed we realised that the interference pattern of BECs released from a double well potential is strongly affected by both the current offset and current turn off time. If these parameters are not well balanced the clouds can shoot off in opposite directions without overlapping in free fall. In the experiment the two Z-wires for the IP trap are connected in series and the wire current is adjusted to vanish within $80\,\mu\text{s}$ as the measurement in figure 3.7 shows.

We use this same design of the controller for all tunable currents as it is flexible and can be

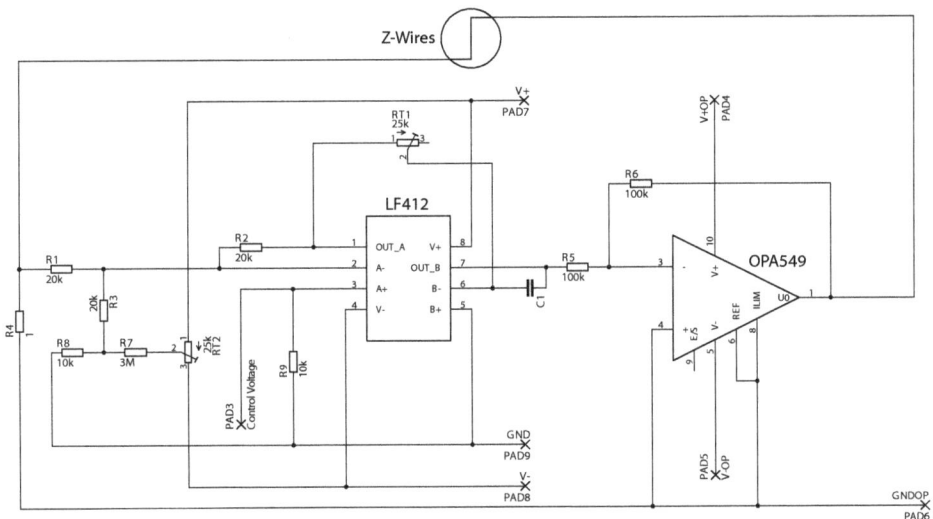

Figure 3.6: Schematic of our home-made current controller. The output current is set by the external control voltage. The LF 421 contains two op-amps comparing the control voltage with a reference. The output is provided by an high power op-amp of the series OPA549. The current running through the chip wires is fed back to the system via the sense resistor R4. The voltage drop over the sense resistor defines the reference voltage and contains the information about the current running through the chip wires.

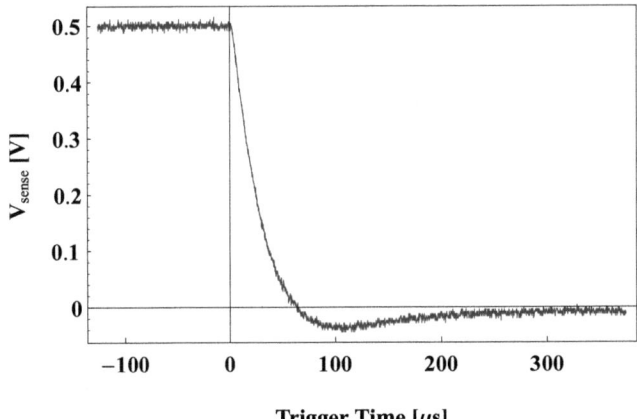

Figure 3.7: The turn off of the wire current. The graph shows the voltage drop over the sense resistor in series with the chip wires. Around $80\,\mu s$ after the turn off the voltage reaches 0 V but overshoots slightly. Both the turn off time and the overshoot is set by the integrator gain. The voltage offset with respect to 0 V can be adjusted with another potentiometer in the circuit. In order to achieve a smooth release of the atoms from the magnetic trap we balance the different parameters carefully.

operated in a bipolar mode. Some shim field coils are operated with fixed currents switched by a digital line. A level setter allows the adjustment of the output current. In these cases we use an assortment of controllers like this and current controllers where the power op-amp is replaced by a power FET.

3.2.7 Radio-Frequency Coupling

Radio-frequency magnetic fields are used at various points in this experiment. The most important tasks are the implementation of forced evaporative cooling and dressed adiabatic potentials. To create rf fields we use two different configurations. One setup runs a rf current through the copper-H below the chip, while the other directly couples the rf current onto the chip wires.

Rf Current in the Copper-H

For evaporative cooling and spectroscopy on the trapped atoms we apply an rf voltage across diagonally separated ends (Z-configuration) of the copper-H to create vertically polarized rf field at the site of the atoms. The rf signal is generated by a Versatile Frequency Generator (VFG150) which allows arbitrary control of an output sequence at frequencies up to 150 MHz. The shape of the sequence, its frequency, amplitude and phase at any point in time is programmed by software written in C++ [84], and downloaded in advance through a USB port to the VFG150. A TTL signal triggers the start of the sequence. The VFG150 output power ranging −69 dBm to 0 dBm, is not sufficient for all our applications and the rf amplifier ZHL-3A from Mini-Circuits provides an additional gain of 25 dB. The amplifier is powered by a 24 V DC low-noise supply for smooth operation. The amplifier output port is then directly connected to the feedthroughs of the copper-H.

Rf Coupling onto the Chip Wires

For splitting the BEC trap into a double well, we use an rf field generated by rf currents in the two Z-wires. These are supplied independently so that we can have full control of the double well potential, as we discuss in the next chapter.

The rf sources are two phase locked DS345 Function Generators from Stanford Research. The maximum output power is 24 dBm at frequencies up to 30 MHz. The DS345 has an amplitude modulation input which allows control of the output amplitude with a voltage between 0 V and 5 V. Each of the function generators is followed by a ZHL-32A amplifier and a ZASWA-2-50DR switch from Mini-Circuits. The switch has the important role of ensuring the rf currents switch on and off with a typical rise/fall time of 5 ns and an in-out isolation of 100 dB at frequencies of up to 100 MHz. The fast switching allows precise timing control. If the output of the DS345 is turned off by applying a modulation voltage of 0 V, we can still observe a background signal at its output port. When the switch is also turned off, its high isolation suppresses this background signal on the chip wires.

We couple the output from the switches to wires using bias-T configuration. The rf signal is coupled via a T1-1T rf transformer from Mini-Circuits to block any DC offset on the output of the switch. Conversely, to prevent a DC current running through the transformer, it is connected in series with a 15 nF capacitor. The value of the capacitor sets the minimum frequency of the rf signal which is coupled to the chip wires to around 200 kHz.

3.3 Making BECs

3.3.1 Computer Control

The experimental cycle necessitates switching currents and laser light with sub-millisecond timing accuracy in a complex sequence. In some cases a simple TTL signal is enough, i.e. to switch a shutter, but some tasks require a sequence of swept control voltages.

Our control computer is based on a 2.08 GHz AMD Athlon XP2600+ processor with 1 GB RAM and five PCI slots. It is equipped with a digital pattern generator PCI-6543 from National Instruments to output the TTL signals. The analogue signals are provided by an analogue output board PCI-6713 as well from National Instruments.

The output is controlled by custom-made Java software called Hermes which allows the specification of an experimental sequence, calculates a waveform and puts it onto the National Instrument cards. The internal timing of the software has a resolution of $200\,\mu s$. Details on the software and the timing can be found in the programmer's thesis [79].

We extended the software to allow automated repetition of identical experimental runs, which reduced the time for data taking dramatically. The basic requirement for that is to automate the acquisition and saving of data and the processing in Mathematica. The communcation between Java and Mathmatica is realized with the J/Link package. After each run Hermes calls a Mathematica function which processes the images and stores experimental parameters in a central database. The database contains all experimental parameters, images and fits and is easily accessible via a web browser or the Mathematica software.

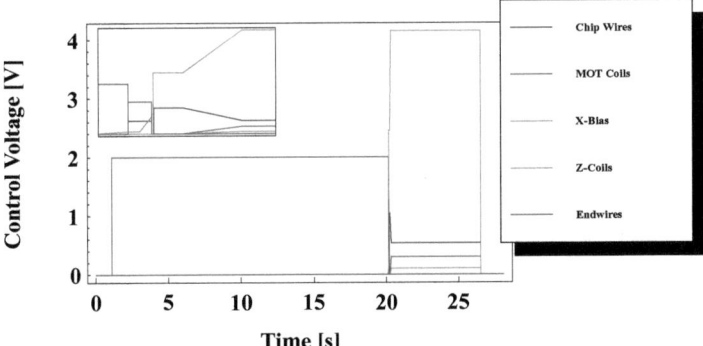

Figure 3.8: Output sequence of the analogue output board for making a BEC. After around 20 s the MOT coils turn off and the magnetic trap is switched on. The whole experimental cycle lasts for around 30 s. The inset magnifies a time interval of 310 ms where the atoms are transfered from the MOT into the magnetic trap.

3.3.2 Experiment Cycle

Here we will describe the standard procedure for making BECs with the presented setup. The magnetically trapped BEC is the starting point for all experiments throughout this thesis. Thus it is necessary at least briefly, to present the experimental cycle at this point, although a detailed description can be found in R. J. Sewell's thesis [78].

One experimental cycle lasts for about 30 s. The example in figure 3.8 shows the behaviour of the control voltages generated by the analogue output board for making a BEC. After the sequence has finished we allow a 30 s break for processing data and cooling down of the coils. Since the process is automated, we analyse the outcome of an experiment about every minute.

Loading the MOT

We start by running a current of 4.5 A through the dispensers and collecting the Rb atoms in the LVIS. The LVIS is connected with the main chamber via a small pinhole and loads a MOT around 4 mm below the atom chip surface. The MOT is created by four laser beams, two counterpropagating, horizontal ones and two counterpropagating beams reflected from the chip surface at an angle of 45° [85]. The trap laser is 2Γ red-detuned from the $F_g = 2 \rightarrow F_e = 2$ transition of the Rubidium D2 line which has the natural linewidth $\Gamma = 6\,\text{MHz}$. The atoms are laser cooled by absorbing photons and randomly reemitting them back into free space. After the photon emission, however, some atoms end up in the $F_g = 1$ state. The repump laser overlapped with the trapping beams transfers the $F_g = 1$ atom back to the $F_e = 2$ state and closes the cooling cycle. The two MOT coils inside the vacuum chamber run a current of 2 A in anti-Helmholtz configuration to provide the quadrupole field required for the MOT with a gradient of 11 Gcm^{-1} along their symmetry axis. After a loading time of 15 s the MOT contains around 10^9 atoms at a temperature of a few hundred μK. We then turn off the LVIS and the dispenser.

UMOT Stage

After waiting for 2 s we move the MOT closer to the chip surface by ramping up an additional homogenous bias field along the x axis within 50 ms which moves the MOT up to 1 mm from the chip surface. Then we turn on a current of 2 A in the trapping wires and 4 A in one of the endwires so that the net current runs in a U-shape. At the same time we turn off the MOT coils while keeping the laser beams on. The net magnetic field forms a quadrupole with large field gradients and the atoms are compressed into a UMOT. The shape of the UMOT is an elongated ellipsoid that matches well the geometry of the magnetic trap. After a hold time of 20 ms we detune the trapping laser to 6Γ and further increase the bias field over another 20 ms, moving the atoms to some 0.5 mm below the chip surface.

Molasses and Optical Pumping

Further cooling is achieved by a molasses process where all magnetic fields are turned off simultaneously. The trapping laser is detuned to 10Γ from the cooling transition. The cloud expands for 3 ms held in place only by the light force.

In the next step we prepare all atoms in the $|F = 2, m_F = 2\rangle$ Zeeman state by an optical pumping cycle of 800 μs. The MOT and repump laser beams are turned off as the X-bias field is ramped up to define a quantization axis for optical pumping by a light pulse of 400 μs. This optical pump beam is 13 MHz red-detuned from the optical pumping transition which is the $F_g = 2 \to F_e = 2$ transition.

Magnetic Trap

The atoms prepared in $|F = 2, m_F = 2\rangle$ of the $5^2S_{1/2}$ electronic ground state are now loaded into a magnetic trap. In order to catch atoms close to the UMOT position, we run a current of 2 A through the Z3 and Z4 wires and overlap the resulting field with a relatively small X-bias field of 13.7 G. The trap frequencies are $\omega_z = 2\pi \times 6\,\text{Hz}$ and $\omega_\perp = 2\pi \times 627\,\text{Hz}$. Due to power dissipaton in the chip, safe operation of the chip wires with a permanent current of 2 A is not guaranteed. After holding the atoms for 50 ms in the initial magnetic trap we therefore linearly ramp the current down to 1 A over 100 ms while ramping up the X-bias field to 23.1 G. At the same time we turn on a bias field of 0.9 G oriented along the z axis in order to prevent atom loss due to spin flips in the trap centre. This step compresses the atoms into the final magnetic trap with trap frequencies of $\omega_z = 2\pi \times 28\,\text{Hz}$ and $\omega_\perp = 2\pi \times 2\,\text{kHz}$, a trap bottom of 630 kHz at a distance to the chip surface of around 160 μm. The adiabatic compression naturally leads to heating, so a 2 s hold time follows so that the hot atoms can spill out of the trap. We end up with 5×10^6 atoms at a temperature of 100 μK.

Evaporative Cooling

In the final phase we cool the atoms down to the BEC transition using forced evaporative cooling. We run an rf current in Z-configuration through the copper-H below the chip wires. This removes hot atoms from the trap through induced spin flips. The rf frequncy is ramped exponentially from initially 15 MHz down to 0.67 MHz. At a value of 1 MHz the rf power is reduced to avoid losses of cold atoms in the BEC phase due to power-broadening. The total ramp lasts for around 4.6 s. After the ramp we let the atoms thermalise for 20 ms and the BEC is ready for experiments. In the next chapter we give details about the BEC itself.

Chapter 4

Characterising the Magnetic Wire Trap

In the last chapter we presented and discussed the basic experimental setup as well as the procedure to achieve BEC in a magnetic trap near the atom chip surface. As we will see later the trapped BEC serves as the input state for the atom interferometer, while the magnetic trap represents the platform for all further experiments. It is necessary to build a more complete picture of the initial system in order to understand the implementation and the operation of the interferometer. In this chapter we characterise the IP trap and the BEC and compare the results with theory.

We will also use the gained knowledge about the magnetic trap to calibrate our imaging system. Taking images is the interface to get any information about the atoms. Absorption imaging, for example, gives us not only information about the density distribution of the atoms but also about the length scale on which the experiment takes place. Additionally to modelling the imaging system, its magnification can also be determined from moving the trap minimum by small modifications to the magnetic fields. At the end of the discussion we will obtain a number for the magnification from an *in situ* measurement.

4.1 The Magnetic Wire Trap

The static magnetic trap is created by running a current through each of two neighbouring, Z-shaped wires. The resulting magnetic field is overlapped with an homogenous magnetic X-bias field in x direction as shown in figure 4.1. At a certain distance above the chip surface the X-bias field cancels the field from the wires and creates a minimum in magnetic potential. In order to avoid atom losses from spin flips a homogenous quantization field B_0 is applied along the z axis. The side pieces of the Z-shape provide the longitudinal confinement which is much weaker than the confinement in radial direction.

4.1.1 Ioffe-Pritchard trap

Such a trapping configuration is known as IP trap and the basic idea is sketched in figure 4.1. The IP field is modelled by

$$\mathbf{B}_{\mathrm{IP}}(r) = B_0 \begin{pmatrix} 0 \\ 0 \\ 1 \end{pmatrix} + B' \begin{pmatrix} x \\ -y \\ 0 \end{pmatrix} + B'' \begin{pmatrix} -xz \\ -yz \\ z^2 - \frac{1}{2}(x^2 + y^2) \end{pmatrix} \quad (4.1)$$

where $B' = \partial |\mathbf{B}|/\partial x = \partial |\mathbf{B}|/\partial y$ and $B'' = \partial^2 |\mathbf{B}|/\partial^2 z$. Near the centre, where the IP trap can be approximated by a cylindrically symmetric, harmonic potential, the trap frequencies in the radial and axial direction are

$$\omega_r = \sqrt{\frac{g_F \mu_B m_F}{m} \frac{B'^2}{B_0}} \quad \text{and} \quad \omega_z = \sqrt{\frac{g_F \mu_B m_F}{m} B''}. \quad (4.2)$$

This kind of trap was first discussed by Ioffe [86] in combination with plasma confinement. Later it was adapted to atom trapping by Pritchard [87, 88]. As is typical the radial trap frequency is much larger than the axial trap frequency $\omega_r \gg \omega_z$. We write the approximate potential $V(\mathbf{r})$ in cylindrical coordinates as

$$V(\mathbf{r}) = \frac{1}{2} m \left(\omega_r^2 r^2 + \omega_z^2 z^2 \right). \quad (4.3)$$

A BEC moving in this potential has an aspect ratio $\lambda = \omega_r/\omega_z$. For large aspect ratios $\lambda \gg 1$ the 1d regime is achieved in the case that both temperature and chemical potential satisfy the condition $T, \mu \ll \hbar \omega_r$.

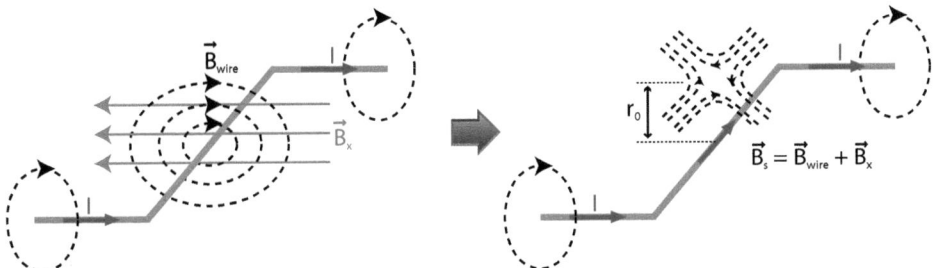

Figure 4.1: Schematic of an IP trap. The field of a Z-shaped wire is overlapped with an external, homogeneous bias field. At a certain height above the wire the magnetic fields pointing in opposite direction cancel each other and form a minimum in magnetic field. Longitudinal confinement is given by the magnetic fields from the side pieces of the Z-shape.

4.1.2 Magnetic Trapping Fields

Infinitely Thin Wire

For a first examination of the magnetic chip trap, let us consider an infinitely long and infinitely thin wire. Biot-Savart's law yields a magnetic field strength at a distance r from the wire

$$B(r) = \frac{\mu_0 I}{2\pi r} \tag{4.4}$$

and by differentiation we find a field gradient of

$$\frac{\partial B(r)}{\partial r} = -\frac{\mu_0 I}{2\pi r^2}. \tag{4.5}$$

Combining the magnetic wire field as sketched in figure 4.1 with a homogenous X-bias field B_x forms a point of zero-magnetic field at a distance

$$r_0 = \frac{\mu_0}{2\pi} \frac{I}{B_x} \tag{4.6}$$

from the wire. The distance is directly proportional to the current I running through the wire. With typical values in our experiment of $I = 2\,\text{A}$ and a bias field strength $B_x = 23.14\,\text{G}$ the distance from the chip surface would be around $173\,\mu\text{m}$.

Finite Size Effects

The geometry of the chip wires, however, is rectangular. The aspect ratio of the wires, with a height of $h = 3\,\mu\text{m}$ and widths of $w = 50\,\mu\text{m}$ and $100\,\mu\text{m}$, is much smaller than unity. Since for typical experiments $h \ll r_0$, the finite height of the wire can be neglected. But the width w is on the order of the separation between trap minimum and chip surface and will therefore play a role in determining the magnetic field. For a long wire with finite width the analytical expression is [89]

$$B(x=0, y) = \frac{\mu_0 I}{\pi w}\left(\frac{\pi}{2} - \arctan\frac{2y}{w}\right). \tag{4.7}$$

The consequence is that at a critical X-Bias field strength $B_{x,\text{crit}} = \mu_0 I/(2w)$ the trap will touch the wire surface. In the limit that the external field is small and hence the distance to the wire is big, equation (4.7) approaches the thin wire result. At a distance $d \sim 170\,\mu\text{m}$ we expect a deviation from equation (4.4) due to finite size effects of less than 5%, and less than 10% for the corresponding field gradient.

A 2d calculation of the two broad-wire configuration in our atom chip experiment is presented in figure 4.2. The origin is located in the middle of the two wire centres which have a separation of $107.5\,\mu\text{m}$. In analogy to our experiment we assume two wires of different width, namely $50\,\mu\text{m}$ and $100\,\mu\text{m}$. With the same parameters we used for the infinitely thin wire, $I = 2\,\text{A}$ (1 A in each wire) and $B_x = 23.14\,\text{G}$, we find a surface to trap minimum distance of $d = 163\,\mu\text{m}$ instead of $173\,\mu\text{m}$.

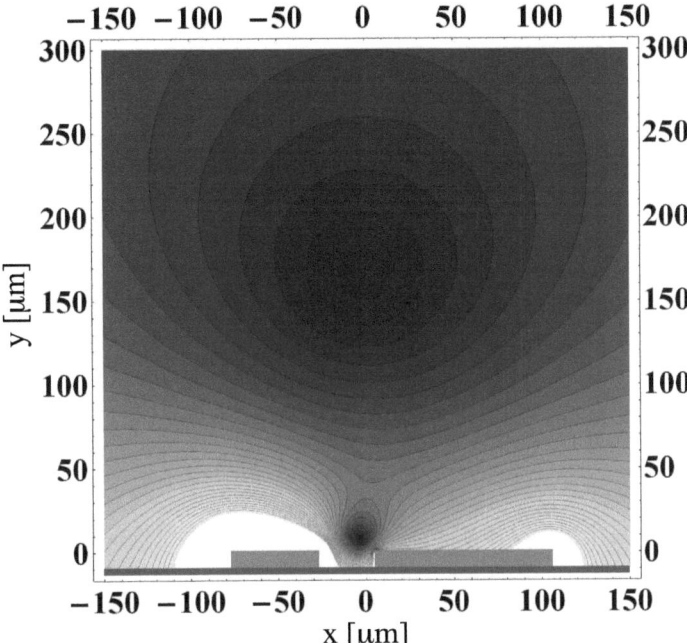

Figure 4.2: The 2d plot shows the magnitude of the magnetic trapping field created by two infinitely long wires and an homogenous external bias field. We use two wires with widths of $50\,\mu m$ and $100\,\mu m$ respectively. The centre to centre separation is $107.5\,\mu m$. The calculation is performed for a current of 1 A in each wire and an bias field strength of 23.14 G. We find two minima forming above the chip surface which cannot be observed in the case of a single wire. The atoms are trapped in the minimum at a distance of around $163\,\mu m$ from the chip surface. This corresponds to a correction of around 5% in comparison to the infinitely thin wire approximation.

4.2 Characterisation of the BEC

Based on the theoretical considerations we create the magnetic trap and fill it with a BEC. The experimental situation is sketched in figure 4.3. The procedure to achieve BEC is as described in section 3.3. Interesting properties which make the trapped BEC distinctive are parameters such as trap frequencies, trap bottom, lifetime, etc. These numbers are important for analysing experiments in later chapters. In particular, the splitting of the BEC into a double well potential directly depends on the parameters of the magnetic trap. We derive a complete picture of the system by analysing step by step the various properties of the trapped BEC.

4.2.1 BEC near the Chip Surface

Figure 4.4 presents the BEC in the magnetic trap at several temperatures. We image the clouds looking along imaging axis 1 after 3 ms time-of-flight. The atom number and temperature of the BEC

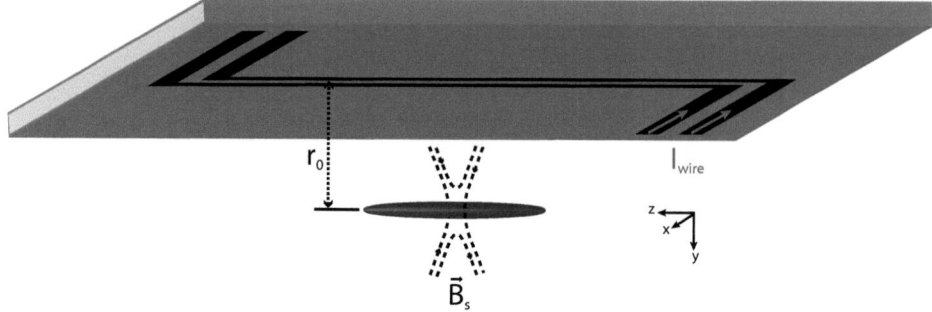

Figure 4.3: Schematic of a trapped BEC near the atom chip surface. The static magnetic field B_s is a result of the chip wire fields and the external bias field. It has the shape of a quadrupole field in the x-y plane with a minimum at a distance r_0 from the chip surface. The BEC is elongated along the cental piece of the Z-wires.

are determined by the stop frequency of the evaporation ramp. The closer we ramp the evaporation stop frequency to the bottom of the trap $g_F\mu_B B_0/h$ the colder and fewer the atoms become. For high stop frequencies three BECs are visible each sitting in a lump of the magnetic trap. The fragmentation of the BECs near current carrying chip wires has repeatedly been reported [90, 91, 92]. It is due to a variation of the magnetic field $\delta B_z(z)$ and hence the trapping potential along the length of the wire. Together with the bias field B_0 the total magnetic field minimum becomes $|B_0 + \delta B_z(z)|$. The BEC will split into different lumps, if the magnitude of the field variation is on the order of the chemical potential $\mu \lesssim \mu_B \delta B_z(z)$.

Ideally the external X-bias field and the magnetic field generated by the current cancel each other to zero at a certain distance from the wire. A local deviation of the current direction, however, causes a tilt between the overlapping magnetic fields and locally an overspill in field strength remains [93]. In the literature three possible reasons for a transverse current component are discussed, namely bulk defects, surface and edge roughness. Latter is the main contribution to fluctuations in the vicinity $r_0 \geq w$. A quantitative analysis on the effect of edge roughness was derived by Wang et al. [94].

Typically we look at the interference pattern of the BECs along the z axis. Fragmentation of the cloud into different lumps leads to a series of independent clouds whose density is integrated along the z axis by the imaging beam. The result is a reduced visibility of the interference pattern. As shown in figure 4.4 we remove atoms from the different lumps by decreasing the stop frequency of the evaporation ramp until we end up with only one single, elongated BEC. In the case that the trap bottoms of the lumps are close to each other, it can happen that the remaining BEC is too small to observe interference after several ms time-of-flight. In that case one could also remove selectively the atoms from unwanted clouds by applying an rf pulse resonant on the trap bottom.

The BEC remaining after the evaporation ramp is the starting point for implementing an atom BEC interferometer. It contains approximately 3×10^4 atoms at a temperature of around $0.5\,\mu$K.

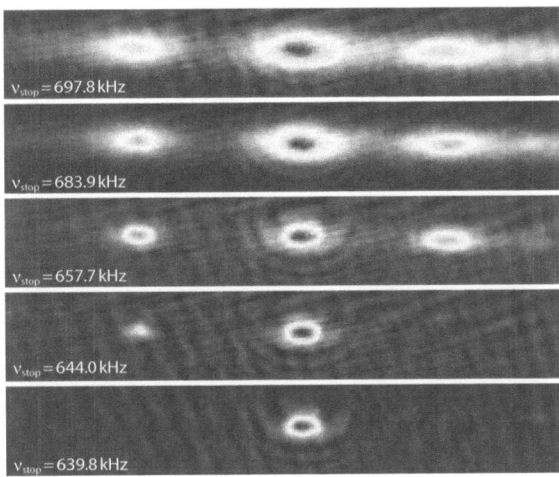

Figure 4.4: Absorption image of cold atom clouds 3 ms after release from the magnetic trap. The final temperature of the atoms is determined by the stop frequency of the evaporation ramp. Due to a variation in a magnetic field along the trapping potential the cloud is fragmented into different lumps. At the transition temperature a BECs form in the different lumps. We remove atoms from the various clouds by further evaportion until all remaining atoms belong to one single BEC.

4.2.2 Trap Frequencies

We determine the longitudinal trap frequency by displacing the BEC and letting it fall back into the initial trap minimum. Increasing the MOT coils current within a few ms shifts the BEC position along the z axis. A sudden jump of the current back to its initial value causes the BEC to oscillate in the magnetic trap. We release the cloud from the trap after various oscillation times and take images after 3 ms time-of-flight on imaging axis 1 to determine its position. The fit in figure 4.5(a) yields an axial trap frequncy of 28.2 ± 0.2 Hz with an e^{-1} damping time of 354 ± 2 ms.

Since the magnetic trap has a large radial confinement, we expect oscillation periods along this direction on the order of the time resolution of our control software. Therefore an *in situ* observation of the oscillation is not possible. Instead we use parametric heating. A small AC current running through the central part of the copper-H creates an oscillating magnetic field along the y axis at the position of the atoms. If the frequency of the AC field is resonant with the radial trap frequency atoms are accelerated and start to spill out after a few ms. From absorption images we obtain the optical density for several driving frequencies. As plotted in figure 4.5(b), this shows a broad, asymmetric resonance curve. Due to the asymmetric wire configuration we find two resonances at around 2005 kHz and 2037 kHz by fitting Lorentzians to the two different slopes on each side of the curve.

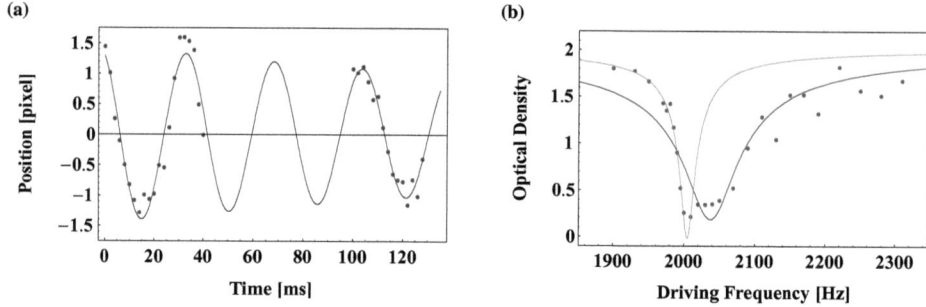

Figure 4.5: (a) The axial trap frequency is much smaller than the radial frequency and can be measured by observing oscillations. Increasing the MOT coils field we displace the BEC and let it fall back into the initial trap. The oscillation in position has a frequency of 28.2 Hz. (b) The radial trap frequency is measured by parametric heating. The minimum in the frequency scan is broad and asymmetric. We fit a Lorentzian on each of the two slopes. The two resonances at 2005 Hz and 2037 Hz are due to the asymmetric wire configuration.

4.2.3 Trap Bottom

The trap bottom defines the energy minimum in the magnetic trap. In general we assume that for a BEC all atoms occupy the lowest energy state. An rf field with a frequency resonant to the trap bottom $\omega_{\rm rf} = g_F \mu_B B_0/\hbar$ [87] couples the five equally spaced energy levels of an $F = 2$ atom and results in spin flips. Atoms driven into an untrapped m_F state leave the trap.

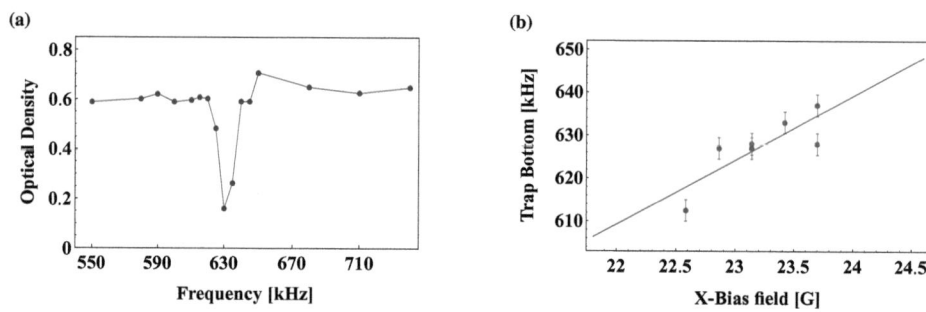

Figure 4.6: (a) A scan of the rf knife frequency performed on our standard magnetic trap. An interaction time of 200 ms at a very low rf power yields a sharp peak at 630 kHz. We run experiments typically at a X-Bias field strength of 23.1 G. (b) The trap bottom changes with the X-Bias field. A higher X-Bias field moves the BEC closer to the chip surface.

We measure the trap bottom by applying a fixed frequency rf knife after the BEC is already prepared. We repeat the experiment at different rf frequencies and determine the optical density after 2 ms time-of-flight. Sharp resonance peaks are achieved at small rf power and long rf knife pulses of 200 ms length. The typical frequency scan in figure 4.6 suggests a trap bottom of 630 kHz or 0.9 G for our standard magnetic trap with an X-Bias field strength of 23.14 G. The distance of the BEC from the chip surface can be controlled with the X-Bias field. A variation in height also changes the trap bottom.

We found that the trap bottom has day to day fluctuations of ±5 kHz but is stable within a series of experiments over one day with a change less than ±1 kHz. Over night we turn off all power supplies. This seems to cause the variation which can be due to a change in the earthing potentials. The consequence is that the current either running through the chip wire or the X-Bias coils is different which influences both the position of the BEC and the trap bottom.

4.2.4 Trap Lifetime

An important property of the BEC is its lifetime in the magnetic trap, since this will limit the maximum available time for experiments. Loss of trapped atoms is caused by background collisions and heating due to noise in the apparatus. For BEC experiments an UHV environment is essential. With a base pressure of $\approx 10^{-11}$ Torr we expect condensate lifetimes of several seconds and also spin flips are suppressed by the quantization field [95].

In order to measure the trap lifetime we hold the atoms inside the magnetic trap and vary the time after which we release the BEC from the trap. The exponential decay in peak optical density is clearly visible in figure 4.7. The fit has an e^{-1} lifetime of around 340 ms which is satisfactory for our purposes. We will see later in this thesis that the limit in available time for a single interferometer measurement is set by other parameters and is in the order of 10% of the magnetic trap lifetime.

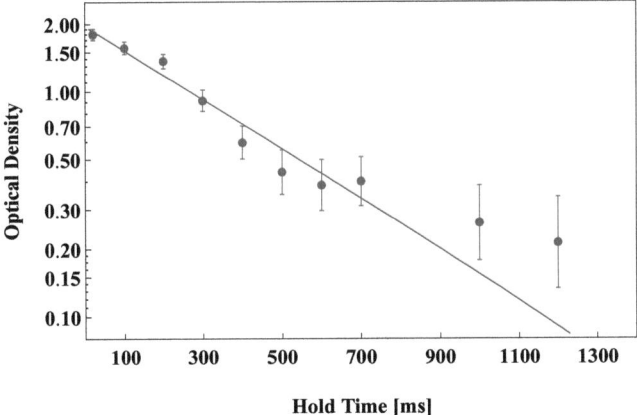

Figure 4.7: Lifetime of the BEC in the magnetic trap. The peak optical density decays exponentially with retention time of the BEC. The measured lifetime of 340 ms, however, is sufficient for the experiments in this thesis.

4.3 Calibration of the Imaging System

All the information extracted from the system comes from camera images. We derive numbers by fitting a model to an image or summing the pixel counts. All parameters related to a length scale are first measured in pixels. If we actually want to measure a physical quantity or make a comparison with a theoretical model a calibration mapping pixels onto a length scale is necessary in most cases. The standard method is to drop the BEC in the gravitational field and deduce the magnification M from its positions after different times-of-flight. This, however, is not acceptable for an apparatus that is supposed to measure the gravitational acclertion g. We therefore present a different calibration method by moving the trapped BEC with respect to the chip surface with magnetic fields. In order to verify the approach we compare it with the standard procedure. The calibration is presented for imaging axis 2 which plays the important role for the interferometer.

4.3.1 Magnification from Gravity

We prepare a typical BEC and release it from the trap. We monitor its vertical position for different time-of-flights and fit the well known model for a uniformly accelerated object to the data. We assume a homogenous gravitational acceleration of $g = 9.81$ m/s^2. Considering the pixel size of 6.8 μm the fit in figure 4.8 yields a magnification of 1.674 ± 0.14. Furthermore, the fit yields a small initial velocity of 6.5 mm/s in $-y$ direction. Fitting the time-of-flight data with the constraint of zero initial velocity delivers a magnification of 1.61 ± 0.03. Since the chip wire and X-Bias turn off is not infinitely fast, the atoms can experience a small kick during release and a velocity offset is most likely.

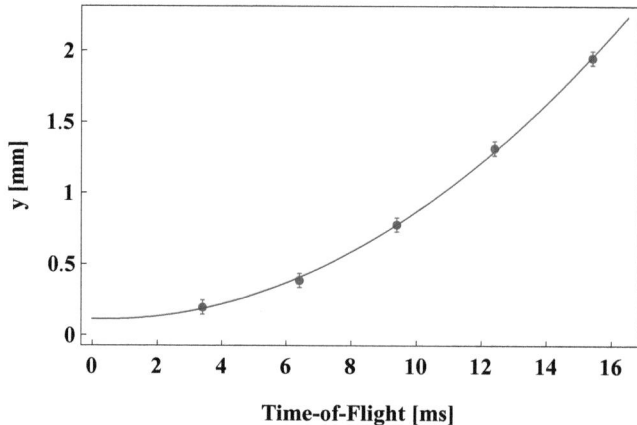

Figure 4.8: Vertical cloud position vs. time-of-flight. The solid line is a fit of a steadily accelerated object which yields a magnification of 1.674 ± 0.14 on imaging axis 2.

4.3.2 Calibration with Magnetic Fields

The second method to determine the magnification of the imaging system is to compare the in-trap position of the BEC with the expectation from the magnetic trap model. Due to diffraction of the imaging beam and improper focussing on the chip surface the absolute distance of the cloud from the atom chip cannot be measured from the image. Small position movements, however, are clearly observable. Modifications of the X-bias field changes the distance between cloud and chip surface. For small changes in the field we make a Taylor expansion of the simple model regarding an infinite long wire of equation (4.6) which yields

$$y(B_x) = \frac{\mu_0 I}{2\pi B_{x,0}} - \frac{\mu_0 I}{2\pi B_{x,0}^2}(B_x - B_{x,0}) + \frac{\mu_0 I}{2\pi B_{x,0}^3}(B_x - B_{x,0})^2 + O\left[(B_x - B_{x,0})^3\right]. \quad (4.8)$$

The Taylor series delivers an analytical expression allowing direct comparison with data points from the experiment.

Therefore we prepare the BEC for different X-Bias values B_x while keeping the chip wire currents constant and image the atoms without time-of-flight inside the trap. In the absorption images we observe a movement of the in-trap BEC along the y axis. We make a precise measurement of the current I running through the chip wires by probing the voltage drop over the sense resistor of the current driver. For magnetic trapping we measure a current of 2.09 A. The X-Bias field was mapped inside the chamber with a Hall probe when building the experiment. Typically we create the BEC at an X-Bias field strength of $B_{x,0} = 23.14$ G. Small changes of the bias field are applied by modification of the corresponding control voltage in the computer output.

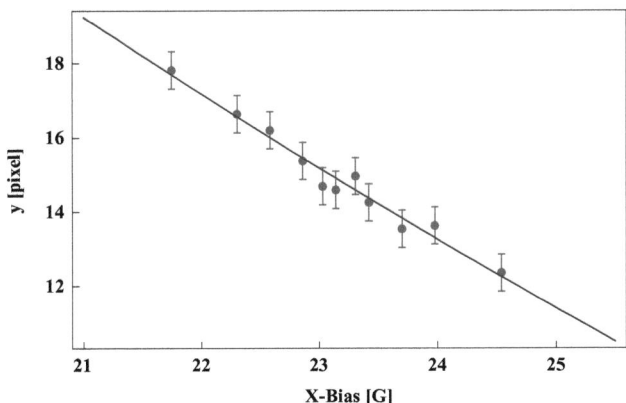

Figure 4.9: Position of the in-trap cloud against X-bias field. For small modifications in the X-bias field the change in height can be linearly approximated. The solid line is the fitted Taylor approximation of the simple model.

We choose a certain region of interest on the absorption images and plot the y position against the bias field in figure 4.9. Calculating the expected absolute height of the cloud and fitting the Taylor series to the data yields a magnification of around 1.70. The spread of the data points around the fit causes an error of 6% in good agreement with the previously described calibration method. To improve our measurement we extend our model by taking finite size effects of the wire into account. We calculate the expected height movement of the BEC analogue to the simulation presented in figure 4.2. The comparison with the data then yields a magnification for imaging axis 2 of 1.67.

The demonstrated procedure represents an alternative method to determine the magnification of an imaging system independent of g. The uncertainty of around 6% is comparable to the error in the g dependent method. As expected the correction for finite size effects makes a small modification to the result. Additional confirmation for the measured magnification we gain from modelling our imaging system presented in figure 3.5 using ABCD matrices [96] and OSLO [97]. The simulation taking the surfaces of the chamber windows, the waveplates, the cube and the lenses into account yields a result of $M = 1.68 \pm 0.01$.

The experimental measurements and the theoretical prediction agree perfectly and we therefore assume a value of $M = 1.67$ as the standard magnification on imaging axis 2 in all following experiments.

Chapter 5

Symmetric and Asymmetric Double Well Potential

5.1 Introduction

The atom chip provides a platform to implement microscopic magnetic traps. Its designs can be adapted especially for a certain task by a particular wire pattern in the surface or by adding micro-optical devices such as fibres. In recent years the flexibility of the atom chip was greatly enhanced by application of rf adiabatic potentials [24]. These new traps opened the path for building double well atom interferometers [13, 23], the implementation of beam splitters [98] and studies of phase fluctuations in low dimensions [99]. The big advantage of dressed state potentials is that they are fully adjustable by the rf current running through the chip. Modern rf generators offer easy control of the various parameters such as frequency and rf power making adjustments in the trap geometry accessible. We in particular use rf adiabatic potentials to split a BEC into a double well and change the relative energy difference between the wells. This flexibility allows more complicated schemes suitable for interferometry.

In this chapter we discuss the basic theory of rf adiabatic potentials and show how the double well potential is implemented and controlled in our experiment. After a theoretical overview in section 5.2 we start in section 5.3 with the splitting of the BEC into two symmetric wells. We describe the scheme used for splitting and how we balance the double well potential. Since the goal for our interferometer is to measure the gravitational energy difference between two BECs, we develop a method to tilt the double well potential with respect to the horizontal axis. Depending on the orientation of the splitting axis a energy gap is established between the two wells. But tilting not only changes the relative height and hence the gravitational potential of the two atom clouds, it also leads to a change in the local magnetic landscape. An accurate characterization of the asymmetric double well is essential for the comparison with later results. We measure the tilt angle and hence the gravitational potential difference. Spectroscopy of the dressed atoms is used to determine the magnetic energy gap. The detailed discussion of the asymmetric double well is presented in section 5.4.

5.2 RF-Adiabatic Potentials

The basic principle of magnetic trapping is the Zeeman effect [100] which shifts the energy of a given spin state $|m_F\rangle$ accordingly to the local magnetic field. A coupling of different m_F states was first suggested for rf evaporative cooling [101] and used in the 1990s to achieve BEC [30, 102, 103]. Instead of using rf fields to spill atoms out of a magnetic trap a new idea by Zobay and Garraway [21, 22] was to dress the Zeeman states in order to create completely new trap geometries. Figure 5.1 visualises the basic principle with the help of a 1d harmonic, magnetic trap. An applied rf field is resonant in two points in space, left and right of the trap minimum. At resonance the avoided level crossing causes a new shape of the trapping potentials. The combination of spin states and rf photons leads to the so-called dressed states as the new eigenstates of the system, labelled by $\tilde{m}_F$ in analogy to the Zeeman states. A double well geometry is implemented by positive $\tilde{m}_F$ states whereas negative values describe potentials used for evaporative cooling. The dressed atom picture was first suggested in the 1960s when Haroche and Cohen-Tannoudji [104, 105] developed a full quantum mechanical treatment.

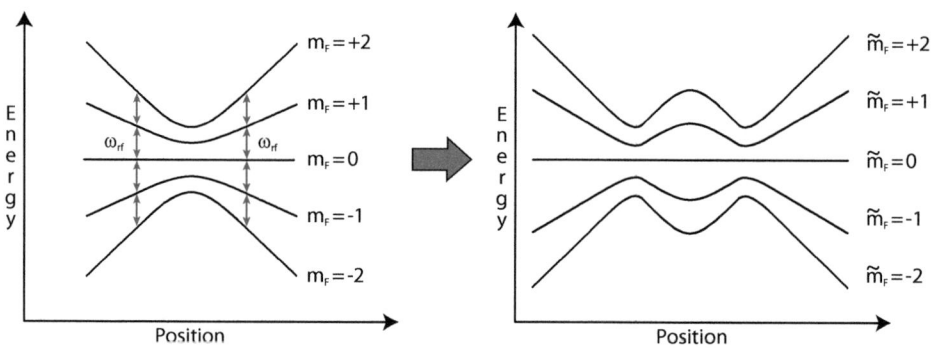

Figure 5.1: Coupling between different m_F states of atoms moving in a 1d harmonic, magnetic trap. The rf field with frequency ω_{rf} is resonant to the energy gap between the states at two points in space. At the resonance the are coupled to each other and the atoms experience a new trap geometry, the so-called dressed state which are labeled with $\tilde{m}_F$. The states with negative $\tilde{m}_F$ are used for evaporative cooling.

5.2.1 The dressed state Hamiltonian

Considering an atom moving in a static magnetic field $\mathbf{B}_s(\mathbf{r})$ dressed by an oscillating field $\mathbf{B}_{rf}(\mathbf{r}, t)$ of frequency ω_{rf} we write down the full Hamiltonian according to Haroche and Cohen-Tannoudji as

$$\hat{H} = g_F \mu_B \left[\mathbf{B}_s(\mathbf{r}) + \mathbf{B}_{rf}(\mathbf{r}, t)\right] \cdot \hat{\mathbf{F}} + \hbar \omega_{rf} \hat{a}^\dagger \hat{a} \tag{5.1}$$

where g_F is the Landé factor, μ_B the Bohr magneton and $\hat{a}$ and $\hat{a}^\dagger$ are the annihilation and creation operator of the rf field. The first part of the Hamiltonian describes the interaction between the external magnetic fields and the atomic angular momentum $\mathbf{F}$. The second part takes the energy of the rf field into account with N the number of photons.

We choose for any fixed point in space $\mathbf{r}$ a coordinate system such that the z axis is aligned with $\mathbf{B}_s$. The component of the oscillating field $\mathbf{B}_{rf}$ orthogonal to the static magnetic field defines the x axis. The rf field with a sinusoidal time dependence is then written as

$$\mathbf{B}_{rf}(t) = \frac{1}{2} \begin{pmatrix} 0 \\ b_{rf,y} \\ b_{rf,z} \end{pmatrix} e^{i\omega_{rf}t} + \text{c.c.}. \tag{5.2}$$

This choice does not describe the general case. In fact we restrict ourselves to this particular one with vanishing x component because it represents our experimental situation as we will see later.

We write equation (5.1) in terms of the bare state basis $|m_F, N\rangle = |m_F\rangle \otimes |N\rangle$ with $m_F = -F, ..., F$, where

$$\hat{H} = g_F \mu_B B_s \hat{F}_z + \hbar \omega_{rf} \hat{a}^\dagger \hat{a} + \frac{g_F \mu_B}{2\sqrt{\langle N \rangle}} \left[\left(B_{rf,y}(t) \hat{a}^\dagger \hat{F}_y + \text{h.c.} \right) + \left(B_{rf,z}(t) \hat{a}^\dagger \hat{F}_z + \text{h.c.} \right) \right]. \tag{5.3}$$

Note that $B_{rf,y}(t)$ denotes the y component of $\mathbf{B}_{rf}(t)$. The resulting Hamiltonian can be split into four parts $\hat{H} = \hat{H}_{spin} + \hat{H}_{field} + \hat{H}_y + \hat{H}_z$. The part of the Hamiltonian describing the interaction between the atom spin state and the rf field splits into two terms $\hat{H}_y$ and $\hat{H}_z$. The first term $B_{rf,y}(t) \hat{a}^\dagger \hat{F}_y$ (+h.c.) raises (lowers) the number of rf photons by one and lowers (raises) the atomic spin state, while the second one containing $B_{rf,z}(t) \hat{a}^\dagger \hat{F}_z$ (+h.c.) raises (lowers) the photon number but leaves the spin unchanged. Therefore a photon with σ^+ polarisation (relative to the local quantisation axis) interacts with the atoms by transfering its spin, while σ^- and π polarised photons do not in general couple to atomic spin states.

We assume that the average number of photons $\langle N \rangle$ is large and the variation around the mean value is small which justifies the approximation $\sqrt{N} \approx \sqrt{N+1}$. Remembering also the relation for the ladder operators $\hat{F}_\pm = \hat{F}_x \pm i\hat{F}_y$ which have the eigenvalues

$$\hat{F}_\pm |F, m_F\rangle = \sqrt{F(F+1) - m_F(m_F \pm 1)} |F, m_F \pm 1\rangle \tag{5.4}$$

we can easily calculate the matrix elements of the Hamiltonian

$$\langle m', N' | \hat{H}_{spin} | m, N \rangle = g_F \mu_B B_s m \delta_{m',m} \delta_{N',N} \tag{5.5}$$

$$\langle m', N' | \hat{H}_{field} | m, N \rangle = \hbar \omega_{rf} N \delta_{m',m} \delta_{N',N} \tag{5.6}$$

$$\langle m', N' | \hat{H}_y | m, N \rangle = \frac{g_F \mu_B}{2} B_{rf,y}(t) \sqrt{F(F+1) - m(m+1)} \delta_{m',m+1} \delta_{N',N-1}$$
$$+ \frac{g_F \mu_B}{2} B^*_{rf,y}(t) \sqrt{F(F+1) - m(m-1)} \delta_{m',m-1} \delta_{N',N+1} \tag{5.7}$$

$$\langle m', N' | \hat{H}_z | m, N \rangle = \frac{g_F \mu_B}{2} B_{rf,z}(t) m \delta_{m',m} \delta_{N',N+1} + \frac{g_F \mu_B}{2} B^*_{rf,z}(t) m \delta_{m',m} \delta_{N',N-1} \tag{5.8}$$

In our basis the matrices H_{spin} and H_{field} are diagonal, while the ones describing the coupling between the states are off-diagonal. The Kronecker delta $\delta_{i,j}$ with $i,j \in \mathbb{N}$ defines the selection rules for transitions between different states in the off-diagonal terms. Equations (5.5) to (5.8) hold as long as the adiabaticity criterion is fulfilled, which means that the orientation of the atomic spin vector can follow the magnetic field vector.

Since the field of the static magnetic trap varies spatially, the Hamiltonian matrix has to be diagonalised at every point in space in order to determine its eigenenergies. The problem can be solved numerically but we gain a better understanding of the atom-field system by introducing the so-called rotating wave approximation (RWA). We discuss the approximation in the following which delivers simple analytical equations valid for certain regimes.

5.2.2 The Rotating Wave Approximation

Following [106] we group the states $\{|m_F, N\rangle\}$ into the manifold $\{|m_F, \kappa - \text{sgn}(g_F) m_F\rangle\}$ denoted by the quantum number κ. For fixed κ this trick delivers matrix elements which are now independent of the photon number N. The next step is to introduce the RWA where an operation of the form $\hat{R}_z(\omega_{\text{rf}} t) = \exp(-i\hat{F}_z \omega_{\text{rf}} t)$ applied to the Hamiltonian transfers it into a rotating frame [107]. The approximation then consists of neglecting all time dependent terms and is only valid for small detunings namely $\omega_{\text{rf}} - \omega_0 \ll \omega_0$, where ω_0 denotes the Larmor frequency. The transformed matrix elements $\tilde{H}_{m'm} = \langle m', \kappa - \text{sgn}(g_F) m' | \tilde{H} | m, \kappa - \text{sgn}(g_F) m \rangle$ are then written as

$$\tilde{H}_{m'm} = g_F \mu_B \left(\frac{\hbar \omega_{\text{rf}}}{g_F \mu_B} - B_s \right) \delta_{m',m} + \kappa \hbar \omega_{\text{rf}} \delta_{m',m}$$
$$- g_F \mu_B \frac{b_{\text{rf},y}}{4} \sqrt{F(F+1) - m(m+1)} \delta_{m',m+1}$$
$$+ g_F \mu_B \frac{b_{\text{rf},y}}{4} \sqrt{F(F+1) - m(m-1)} \delta_{m',m-1}. \quad (5.9)$$

Since we are not interested in the state of the rf field we drop the rf photon energy $\kappa \hbar \omega_{\text{rf}}$ which couples the different manifolds of rf photons. The comparison then shows that the matrix elements $\tilde{H}_{m'm}$ are equal to that of a spin particle in a static magnetic field written as

$$\mathbf{B}_{\text{eff}}(\mathbf{r}) = \left(B_s - \frac{\hbar \omega_{\text{rf}}}{g_F \mu_B} \right) \mathbf{e}_z + \frac{b_{\text{rf},y}}{2} \mathbf{e}_y. \quad (5.10)$$

Defining $\Omega_\theta(\mathbf{r}) = g_F \mu_B |\mathbf{B}_{\text{eff}}(\mathbf{r})|$, the Hamiltonian of the simplified system is just

$$\hat{H}_{\text{RWA}} = \Omega_\theta(\mathbf{r}) \hat{F}_\theta. \quad (5.11)$$

which is diagonal in the dressed state basis $|\tilde{m}_F\rangle$ where $\tilde{m}_F = -F,...,F$ and $\hat{F}_\theta = \cos\theta \hat{F}_z + \sin\theta \hat{F}_y$ is the spin operator in the new basis with the angle θ as defined below. The dressed state can be interpreted as the magnetic quantum number with respect to the quantization axis defined by the effective magnetic field. If the adiabaticity criterion is satisfied the position dependent potential becomes

$$U(\mathbf{r}) = \tilde{m}_F \hbar \sqrt{\delta^2 + \Omega^2}, \tag{5.12}$$

where δ^2 is called the resonance term and depends on the local detuning between the Larmor frequency and the rf field

$$\delta = \frac{|g_F \mu_B|}{\hbar} |\mathbf{B}_s(\mathbf{r})| - \omega_{\text{rf}}. \tag{5.13}$$

The second term Ω^2 is the so-called coupling term. It is position dependent and takes the influence of the rf field into account which depends also on its local orientation with respect to the static magnetic field. In our case of linearly polarized light the coupling term becomes

$$\Omega = \frac{|g_F \mu_B|}{\hbar} \frac{|\mathbf{b}_{\text{rf}}(\mathbf{r}) \times \mathbf{B}_s(\mathbf{r})|}{2 |\mathbf{B}_s(\mathbf{r})|}. \tag{5.14}$$

For a static magnetic field perpendicular to the rf vector the coupling term becomes $\Omega = \frac{|g_F \mu_B|}{2\hbar} |\mathbf{b}_{\text{rf}}(\mathbf{r})|$ which is in general called the Rabi frequency. Considering equation (5.10) the effective magnetic field lies in the x-z plane at an angle

$$\tan \theta = -\frac{\Omega}{\delta} \quad 0 \leq \theta \leq \pi \tag{5.15}$$

to the quantization field. We can therefore describe the dressing of the atoms as a tilt of the initial spin vector and the dressed states are written in terms of a rotation operator $\hat{R}_x(\theta) = \exp(i\hat{F}_x \theta)$

$$|\tilde{m}_F\rangle = \hat{R}_x(\theta) |m_F\rangle. \tag{5.16}$$

A sudden turn off of the rf field causes the atoms to be projected onto their bare magnetic spin states. Thus, instead of obtaining the initial state in which the atoms were prepared they can rather end up in a superposition depending on $R_x(\theta)$.

5.2.3 Double Well Spectroscopy

We have now achieved a concrete picture of the dressed adiabatic potentials, given in the RWA by equation (5.12). With the right experimental parameters equation (5.12) describes a double well potential. In general we determine the trap bottom of a magnetic trap with spectroscopy using an rf field. For the double well potential we expect that its spectrum is given by $U(\mathbf{r})$. In the following, however, we will see that the formal derivation of this result is not that simple in the vicinity of another rf field.

Doubly Dressed Hamiltonian

The behaviour of spectroscopy on dressed adiabatic potential can be understood by solving the dressed atom Hamiltonian with an additional, weak rf field of frequency ω_{sp}, field strength b_{sp} and the respective Rabi frequency $\Omega_{sp} = g_F \mu_B b_{sp}/2\hbar$. The two rf fields have a frequency difference of $\Delta = \omega_{sp} - \omega_{rf}$. We assume that both rf fields are linearly polarized along the y axis. The new Hamiltonian is extended by the additional rf field and since it is very weak one can then apply the same RWA as described in section 5.2.2 which leads us to [108]

$$\hat{H}(\mathbf{r},t) = \Omega_\theta \hat{F}_\theta + \Omega_{sp}\left[\hat{F}_x \cos(\Delta t) + \hat{F}_y \sin(\Delta t)\right]. \quad (5.17)$$

This Hamiltonian describes the system in a rotating frame of frequency ω_{rf}. The first term describes the dressed adiabatic potential, while the second term is time-dependent. Expressing the Hamiltonian through a rotation $\tilde{H} = \hat{R}_\Delta \hat{H} \hat{R}_\Delta^\dagger$ with $\hat{R}_\Delta = \exp\left(i\hat{F}_\theta |\Delta| t\right)$ permits the application of a second RWA yielding the effective Hamiltonian

$$\tilde{H}(\mathbf{r}) = -(|\Delta| - \Omega_\theta(\mathbf{r}))\hat{F}_\theta + \frac{\Omega_{sp}}{2}\left[1 + \text{sgn}(\Delta)\cos\theta(\mathbf{r})\right]\hat{F}_{\perp\theta}, \quad (5.18)$$

which is equivalent to the Hamiltonian in equation (5.11) with the coupling

$$\Omega_\Delta(\mathbf{r}) = \sqrt{(|\Delta| - \Omega_\theta(\mathbf{r}))^2 + \frac{\Omega_{sp}^2}{4}\left[1 + \text{sgn}(\Delta)\cos\theta(\mathbf{r})\right]^2}. \quad (5.19)$$

The eigenstates of the now doubly dressed states are labeled by $|\tilde{\tilde{m}}_F\rangle$ and correspond to the single dressed states $|\tilde{m}_F\rangle$ rotated by an angle

$$\tan\theta_\Delta = \frac{\Omega_{sp}\left[1 + \text{sgn}(\Delta)\cos\theta\right]}{2(|\Delta| - \Omega_\theta)} \quad (5.20)$$

analogue to the single rf field case. We see that the physics is scarcely more complicated with the application of a second, weak rf field. We turn next to the transition frequencies between the singly dressed states, which are not immediately obvious from the Hamiltonian.

Outcoupling of Atoms

The frequency spacing between the singly dressed states at a certain space point $\mathbf{r}$ is given by $\Omega_\theta(\mathbf{r})$. For large enough detuning Δ we can always find a surface inside the double well potential where the resonance condition $\Omega_\theta(\mathbf{r}) = |\Delta|$ is fulfilled. At these space points atoms are outcoupled with a coupling rate of $\Omega_{sp}\left[1 + \text{sgn}(\Delta)\cos\theta\right]/2$. Indeed we find from equation (5.11) that the resonance condition is realized by the two values $\theta_0 = \arcsin(\Omega/\Delta)$ and $\pi - \theta_0$ within one of the wells. In the literature the two resonances are refered to as inner (IR) and outer resonances (OR). Atoms which are hot enough to reach the IR and the OR make transitions to untrapped states and are removed from the double well potential. This priniciple was successfully implemented for evaporative cooling within rf adiabatic potentials [109].

At an angle $\theta_0 = \pi/2$ the resonances meet in the well centres. For a BEC trapped in a double well potential all atoms occupy the lowest energy state and are therefore located in the well minima. The point is that to remove all atoms with an additional rf field both outcoupling at the well centre and the resonance condition have to be fulfilled which means $\Omega_\Delta (\Omega_\theta = |\Delta|) = \Omega_\Delta (\theta_0 = \pi/2)$. Using the fact $\cos\theta = \sqrt{\Delta^2 - \Omega_\theta^2}/|\Delta|$ which can be derived from equation (5.18) we find the relation

$$\frac{\Omega_{sp}}{2}\left(1 + \frac{\sqrt{\Delta^2 - \Omega_\theta^2}}{|\Delta|}\right) - \sqrt{(\Delta - \Omega_\theta)^2 + \frac{\Omega_{sp}^2}{4}} = 0. \quad (5.21)$$

Therefore we expect resonances of the spectroscopy signal at the frequencies $\omega_{sp} = \omega_{rf} \pm \Omega_\theta$. In a perfectly symmetric double well, all atoms are removed from the trap at two frequencies. In an asymmetric double well each of the wells has its own resonances and the energy difference in the trapping potential is given by the difference of the corresponding transition frequencies.

Beyond the Rotating Wave Approximation

The RWA treatment of the Hamiltonian with two rf fields predicts two allowed transitions. The problem is that the RWA only considers couplings within one manifold of κ as discussed in section 5.2.2. However, for large rf dressing field amplitudes the fluctuation in photon number becomes large and transitions to higher manifolds have to be considered. The occurance of new sideband resonances for Zeeman states due to higher manifolds were first proposed by Haroche [110]. In connection with adiabatic potentials the transition matrix elements for the spectroscopy field were calculated numerically by Hofferberth et al. [106] who found a chain of allowed transition frequencies.

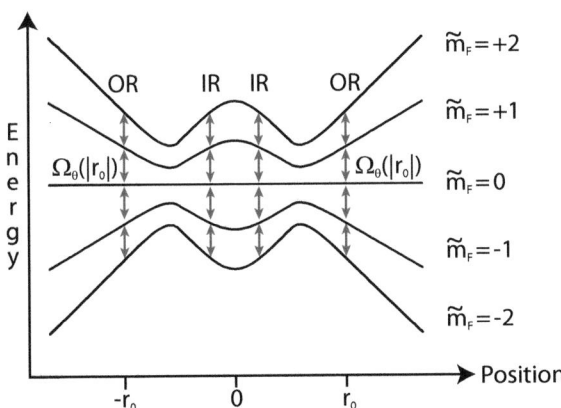

Figure 5.2: Transitions between the various dressed states $\tilde{m}_F$ occur where the spectroscopy field is resonant to the level spacing $\Omega_\theta(r)$. This happens at four surfaces in the dressed potentials. Due to multi photon processes coupling the different manifolds transitions occur at the same positions also for the condition that $\Omega_\theta(r_0) = |\omega_{sp} - n\omega_{rf}|$ with $n \in \mathbb{N}$.

The weak spectroscopy field is intended to induce transitions between the dressed states $\tilde{m}_F$ and $\tilde{m}'_F$. Due to absorption and emission of photons we expect similar selection rules as for the dressing field with $\tilde{m}_F - \tilde{m}'_F = \pm 1$. It is intuitively clear from figure 5.2 that a spectroscopy field of frequency $\omega_{sp} = \Omega_\theta(\mathbf{r})$ drives such a transition at the space point $\mathbf{r}$. Since we assume a strong rf dressing field with a large variation of photons in the order of $\sqrt{\langle N \rangle}$, a coupling between different manifolds can occur. The link between the different manifolds is induced by multi photon processes and hence the resulting transition frequencies are

$$\omega_{sp} = n\omega_{rf} \pm \Omega_\theta(\mathbf{r}) \qquad (5.22)$$

for $n = 1, 2, ...$ and $\omega_{sp} = \Omega_\theta$ for $n = 0$. In the case $n = 1$ with $\omega_{sp} - \omega_{rf} = \Omega_\theta$, for example, an atom undergoes a transition from $\tilde{m}_F = 2$ to $\tilde{m}_F = 1$ by emitting two dressing field photons and absorbing one photon of the weak probing field.

5.3 Splitting of the BEC

We now have enough theoretical background to describe an experimental situation. The goal is to deform a magnetically trapped BEC into a symmetric double condensate. The experimental situation is sketched in figure 5.3. The static currents I_{wire} generate trapping fields for the magnetic trap. In addition, the two chip wires Z3 and Z4 also carry an rf current creating an rf magnetic field at the position of the atoms. Before we discuss the technical details for splitting a BEC, however, we want to have some further thoughts on the configuration of the magnetic fields and the resulting potentials.

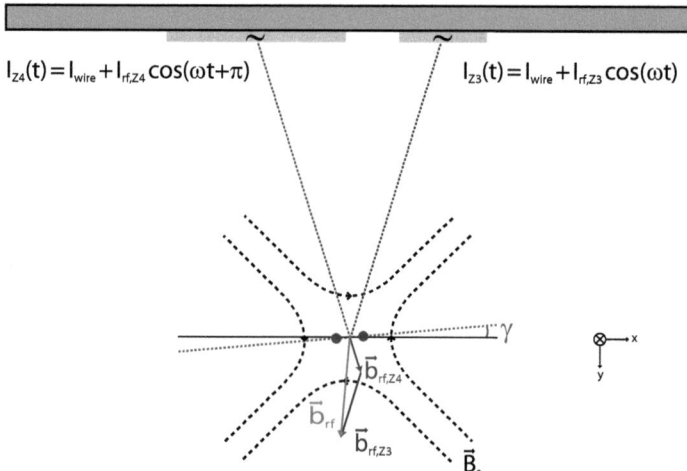

Figure 5.3: The currents are in the two chip wires are a combination of static and rf currents. The static currents provide magnetic fields for the IP trap whereas the rf currents with a π phase shift create a linearly polarized rf magnetic field at the position of the atoms. Due to dressing of the atomic spin states the atoms experience a double well potential in the x-y plane. The direction of splitting encloses a certain angle γ with the horizontal depending on the orientation of the rf amplitude vector $\mathbf{b}_{rf}$.

5.3.1 The Double Well Potential

The magnetic field of the two trapping wires overlapped with a homogenous external magnetic field produces a quadrupole field in the x-y plane. The magnetic field of the IP trap in the $z = 0$ plane is given in polar coordinates by

$$\mathbf{B}_s(\mathbf{r}) = \begin{pmatrix} B'r\cos(\varphi + \pi/4) \\ -B'r\sin(\varphi + \pi/4) \\ B_0 \end{pmatrix}. \tag{5.23}$$

The rotation by $\pi/4$ takes the geometry of our system into account, B' is the magnetic field gradient and B_0 is the homogenous Ioffe field strength. We generalize the rf field vector for linearly polarized light to an arbitrary direction in the x-y plane

$$\mathbf{b}_{rf} = \begin{pmatrix} b_{rf}\cos(\vartheta) \\ b_{rf}\sin(\vartheta) \\ 0 \end{pmatrix}. \tag{5.24}$$

The absolute value of the rf field $|\mathbf{b}_{rf}| = b_{rf}$ is independent of its direction which is defined by the angle ϑ. The explicit description of the magnetic fields permits the calculation of the coupling term Ω in the dressed adiabatic potential of equation (5.12) which becomes

$$U(\mathbf{r}) = \tilde{m}_F \sqrt{[g_F\mu_B|\mathbf{B}_s(\mathbf{r})| - \hbar\omega_{rf}]^2 + \left(\frac{g_F\mu_B b_{rf}}{2|\mathbf{B}_s(\mathbf{r})|}\right)^2 \left[B_0^2 + B'^2 r^2 \sin^2(\vartheta + \varphi + \pi/4)\right]}. \tag{5.25}$$

The direction of the smallest coupling defines the direction of the splitting and is determined from the condition $\varphi + \vartheta + \pi/4 = 0$ or $\varphi + \vartheta + \pi/4 = \pi$. When the rf field is linearly polarized along the y axis, with $\vartheta = \pm\pi/2$, the minima of the coupling term are located at $\varphi = \pm 3\pi/4$ and $\varphi = \pm\pi/4$. Since the sign of the rf field vector is changing in time, the splitting occurs along the x axis. A rotation of the rf field around the z axis results in a rotation of the splitting direction by the same amount but opposite direction.

For a sufficiently large Ioffe field we can make the assumption $r \ll B_0/B'$ and an approximation of equation (5.25) yields the position of the potential minima [111]

$$r_0 = \frac{1}{\sqrt{2}B'}\sqrt{b_{rf}^2 - B_{crit}^2}. \tag{5.26}$$

For creating a double well potential we follow the example of Schumm et al. [13] and use a red-detuned rf signal with $\hbar\omega_{rf} < g_F\mu_B B_0$. The potential splits into two minima only if the rf field strength is larger than the critical field strength

$$B_{crit} = 2\sqrt{B_0 \frac{g_F\mu_B B_0 - \hbar\omega_{rf}}{|g_F\mu_B|}}. \tag{5.27}$$

The transformation from a single well to a double well is hence possible by tuning one of the parameters b_{rf}, B_0 or ω_{rf}. The separation distance, however, is independent of the orientation of the linearly polarized rf field.

Although the rf field is not resonant with the Zeeman energy gap at any point in space, the states are still coupled and it is possible to create two wells with a sufficiently large rf field strength. Since in the trap centre the detuning is nearest to resonance, the potential is pushed up furthest there and results in a local maximum.

5.3.2 The Splitting Process

In the following we describe the procedure to create symmetric double well potentials, used throughout all the experiments of this thesis. The exact numbers for the parameters were found by optimising the experiment for the best results.

The double well potential is implemented by overlapping the DC current in the trapping wire with two independent, phase-locked rf currents. The coupling is realized with a bias-T as described in section 3.2.7. The phase between the two rf currents is π which creates a linearly polarized rf magnetic field at the position of the atoms. The field is polarized along the vertical y axis which yields a splitting along the horizontal. In order to split the BEC into the double well we ramp up the rf currents linearly over 20 ms. The rf frequency is fixed at 540 kHz which is 90 kHz red-detuned from the bottom of the magnetic trap. In experiments we found that we start to outcouple atoms from the trap due to power broadening at detunings smaller than 80 kHz. The final rf field strength is adjustable. For balanced splitting the rf current in each wire is adjusted independently. The ratio of these currents, however, is fixed during the whole process. Estimate from the rotating wave approximation and numerical simulations for our setup show that the BECs are well separated for fields of around 0.8 G [78]. From equation (5.26) we estimate that two minima start to form at around 0.68 G, hence the actual spatial separation happens within the last 5 ms of the splitting process.

5.3.3 Balancing of the Double Well

After ramping up the rf power in the two wires the wells can differ in the energy of the lowest states, causing the two BEC clouds to have unequal atom number. Equation (2.20) shows that the contrast of the fringe pattern is maximal for an equal fraction of atoms in each well. Therefore it is favourable to prepare the atoms in a symmetric double well.

As we have already discussed in detail, the splitting direction depends on the spatial orientation of the rf field. Since the two trapping wires are not symmetric and have different impedances, an equal rf output power on both rf generators does not split the cloud horizontally. Since the rf field gradient of an infinitely long wire drops with distance from the wire $\partial |\mathbf{b}_{rf}|/\partial r \propto 1/r$, the cloud closer to the chip surface sees a larger rf magnetic field and hence has a larger Rabi coupling Ω which increases the trap bottom of the corresponding well. In the case of two broad wires we expect a modified scaling of the rf field gradient, but still the rf field will not be uniform over the extent of the double well. For balanced splitting the contributions from the magnetic interaction potential and gravity have to cancel the energy difference between the wells. Only if

$$\Delta V_{\text{grav}} + \Delta E_{\text{mag}} = 0 \tag{5.28}$$

where ΔE_{mag} is the difference in magnetic field energy and ΔV_{grav} in gravitational potential, a balanced potential can form.

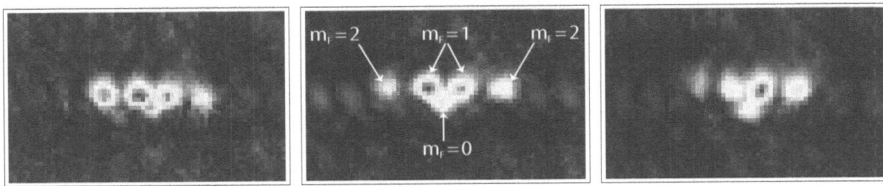

Figure 5.4: Images of the double well atoms after applying the balancing scheme. We turn off the rf field while maintaining the IP trap. The clouds from the two wells projected onto the $m_F = 1$ and 2 magnetic spin states are accelerated inside the harmonic potential and cross each other. After the release from the trap the clouds are thrown in opposite directions. Due to the different magnetic coupling not only the atoms from the two wells are identifiable but also the different spin states. The blob in the middle refers to the $m_F = 0$ state falling in the gravitational potential. The image in the middle shows a balanced double well, while the others have an emphasized left and right well respectively.

A balanced double well potential is achieved by monitoring the atom number in the minima of the potential for experiments with different rf current ratios. The imaging system, however, does not permit the resolution of the double well potential with its separation of typically $2\,\mu$m to $4\,\mu$m. We check that the clouds are of equal size using a scheme desribed in [112]. The BEC is split and instead of simply releasing it, we first turn off the rf field. During the rf turn off the atoms are projected onto their bare magnetic spin states which move now in the single IP trap. The atoms in the high field seeking states with $m_F = -1$ and $m_F = -2$ are immediately expelled from the trap, while the $m_F = 0$ state atoms fall freely under gravity. Atoms in the $m_F = 1$ and $m_F = 2$ state are accelerated from the positions of the double well minima towards the centre of the IP trap. The clouds from the two wells cross in the trap centre without disturbing each other due to their low density. After $400\,\mu$s, which corresponds to approximately three-quarters of a radial oscillation period, we additionally turn off the static magnetic fields and the kinetic energy gained from the movement in the magnetic trap separates the clouds of the two wells far enough from them to be optically imaged. Since the potential is different for the $m_F = 1$ and $m_F = 2$ spin states, both have different kinetic energy and are resolved after 2 ms time-of-flight. The resulting images for wells with different imbalancing are shown in figure 5.4. Counting the atoms in the different components gives an estimate of the size of the BECs inside the adiabatic potential.

The fraction of atoms in each well in dependence of the current ratio is shown figure 5.5. We repeat the balancing experiment for different ratios and count the atoms in the $m_F = 1$ and 2 states. In the presented case the clouds in each well are of equal size at a ratio of $I_{Z4}/I_{Z3} = 0.60$. Increasing the

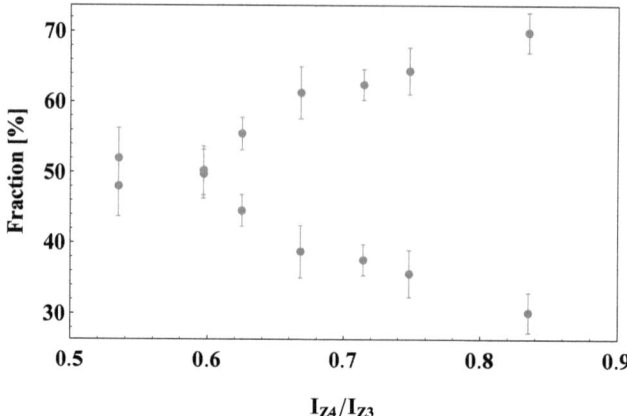

Figure 5.5: Fraction of atoms in each of the well in dependence wire current ratio. The fraction is determined by counting the atoms in the $m_F = 1$ and 2 states using the described balancing scheme. A higher ratio results in an increased atom number in the right well (light grey) while the atom number in the left well (dark grey) is decreased. At a ratio of $I_{Z4}/I_{Z3} = 0.60$ the clouds are well balanced.

rf current in Z4 while decreasing it in the other by the same amount increases (decreases) the number of atoms in the right (left) well and vice versa.

The described scheme is used at various points within this thesis and is succesfully applied to optimise the splitting process. However, it is not a precise measurement of the energy difference in the wells. Figure 5.4 shows balanced clouds at a current ratio of $I_{Z4}/I_{Z3} = 0.65$ which is the value we use for splitting in the following experiments. We realised, however, that for optimal results this value should be checked regularly and modifications in the order of 10% can be necessary on a time scale of a few weeks.

5.4 Implementation and Characterisation of the Asymmetric Double Well

In a symmetric double well potential the trap bottoms of both sites have the same energy level and we call the potential balanced. Especially, in the case of a very slow splitting of a BEC where the system has enough time to adjust itself we end up with two clouds of equal atom number. A deviation from a balanced energy difference between the lowest states in each well leads to a so-called asymmetric double well with unequal atom number on each site. A precise measurement of the gravitational constant g by splitting a BEC into an asymmetric potential and counting the atoms in each well was demonstrated on an atom chip [113] in 2007. However, if we want to apply an energy shift while maintaining an uniform atom distribution, we rather first split the BEC into balanced clouds and then apply an asymmetry afterwards. At large enough separation the sites lose their coupling link and an exchange of atoms is no longer possible. Our goal in this section is to imbalance the energy levels of the two wells in such a way that all the other parameters like atom number, splitting distance, etc. remain unaltered. After developing a scheme for tilting the orientation of the double well potential with respect to the horizontal, the following task is to characterise the energy difference between the wells which is due to a gravitational shift and a shift in Rabi coupling.

5.4.1 Tilting of the Double Well

We determined the ratio of the rf wire currents for a symmetric double well from the balancing experiment. Together with the desired splitting distance this ratio defines the power outputs we dial up on the two rf generators. If we want to implement an asymmetry, that is an energy difference ΔV between the wells, we can just dial up a slightly different rf current ratio. The disadvantage is then that the asymmetry cannot be varied during an experimental cycle and the splitting process is always imbalanced resulting in a difference in atom number between the clouds. Therefore we use a much more flexible scheme.

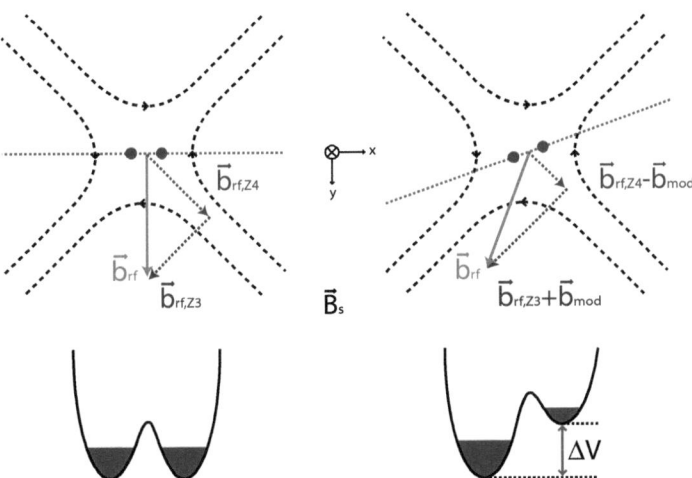

Figure 5.6: Schematic of the tilting process of the double well potential. By adding and subtracting a small modulation field $\mathbf{B}_{mod}$ to the magnetic fields $\mathbf{B}_{rf,Z3}$ and $\mathbf{B}_{rf,Z4}$ respectively, the double well is tilted around the z axis. The resulting height difference introduces an energy difference ΔV and hence an asymmetric potential.

Parametrising the Asymmetry

The rf sources have a modulation input which takes values between 0 V and 5 V corresponding to zero output and maximum output of the dialed power respectively. The amplitude of the output signal is proportional to the modulation voltage. In fact we control the linear rf ramp for the splitting process with the modulation input. First, we adjust all parameters such that we create a balanced double well at a modulation voltage of 4.5 V instead of 5 V. After the splitting process has finished we increase the modulation voltage on one rf source while decreasing it on the other source by the same amount leading to a modification of the rf magnetic field vectors. The situation is illustrated in figure 5.6. The resulting rf field vector is the superposition of the fields generated by both wires and hence tilts by an angle ϑ with respect to the y axis while its length stays constant. According to equation (5.25) this rotates the splitting direction of the double well by the same amount but with opposite sign. In order to parametrise the imbalance we introduce the artificial asymmetry parameter

$$\alpha = 10 \frac{(V_{\text{mod},Z3} - V_{\text{mod},Z4})}{2}, \tag{5.29}$$

where $V_{\text{mod},Z3}$ and $V_{\text{mod},Z4}$ are the modulation voltages on the according rf generators with the condition that

$$V_{\text{mod},Z3} + V_{\text{mod},Z4} = 9 \text{ V}. \tag{5.30}$$

Although the modulation voltages are changed in opposite direction, we have to be aware that in the described scheme the total rf field strength at the space point of the atoms changes by a small amount because the output amplitude of the rf generator is proportional to the selected power. Splitting into a symmetric double well, however, requires an initial imbalance in output power. It follows that by tilting we also change the rf field amplitude by a small amount.

Behaviour of the Well Separation

The splitting distance is related to the fringe wavelength according to equation (2.21) which is a good approximation at large enough well separations (see later discussion). Since an in-trap observation of the splitting distance is impossible with our imaging system, we make use of the fringe wavelength to quantify the effect of the tilting scheme on the well separation. At this point we anticipate the next chapter where we discuss in detail the observation and analysis of an interference pattern. We split the BEC into a symmetric double well potential defined by $\alpha = 0$ V and a power ratio of $P_{Z4}/P_{Z3} = -13.9$ dBm/ -9.9 dBm. We then ramp the modulation voltages to a certain asymmetry α over 1 ms and determine the fringe wavelength Λ after 12.4 ms time-of-flight as described in chapter (6).

By observing the fringe wavelength we indirectly also measure the splitting distance d. In figure 5.7 we plot the wavelength Λ for different double well configurations which clearly exhibits a correlation between well separation and asymmetry parameter. Every data point represents 20 repeated experiments. The change of the quantity Λ can be linearly approximated within the observed range of tilt angles. From the linear regression we are able to determine the variation to within 0.6%. Taking the camera setup and the magnification into account the symmetric potential at $\alpha = 0$ V has a well separation of around 3.23 μm and varies between 3.05 μm and 3.43 μm over a range of $\alpha = \pm 5$ V which corresponds to a modification on the order of $\pm 6\%$. The well separation increases (decreases) in the direction of positive (negative) values of the asymmetry parameters.

This experiment shows qualitatively the behaviour that we expect from the tilt process. The quantitative use of the data, however, is restricted by the experimental parameters. The condition is that

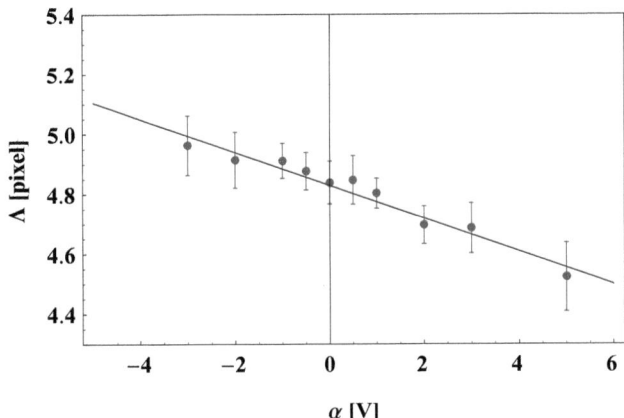

Figure 5.7: Behaviour of the fringe wavelength in dependence of the asymmetry parameter α. Every data point represents the average of 20 repeated experiments. Over the small range of operation a linear approximation seems to be justified. The fit yields a slope of -0.055 ± 0.005 pixel/V.

we perform all experiments concerning the asymmetric well at the specified power ratio and within the studied range of the parameter α. Otherwise we would have to expect a modified behaviour of the splitting distance. Where it is justified, however, the linear fit gives us a simple approximation to describe the double well behaviour.

Observing the Tilt

In order to proof that we are able to tilt the potential we dial up an rf power of $P_{Z4}/P_{Z3} = -12.0\,\text{dBm}/-8.0\,\text{dBm}$ which gives a fairly large splitting distance and release the clouds using the balancing scheme. Already after 4 ms time-of-flight a change in height difference is clearly observable for different α. In figure 5.8 we present averaged images for the tilt parameters $\alpha = -5$ V and $\alpha = 5$ V. The dashed line represents the horizontal with respect to the camera chip. A position change of the atoms in the $m_F = 1$ and $m_F = 2$ is clearly visible. The opposite movement in height is the result of the rotation around the z axis.

Finally, we present a scheme which permits full control of the double well potential. Not only the splitting distance and the balancing can be controlled with the rf currents but also the energy difference can be adjusted. Basically there are two contributions to the energy shift, the difference in gravitational potential and the shift in magnetic energy. Since in the rotated potential one well is closer to the chip surface than the other, the atoms closer to the chip see a much larger Rabi coupling. The characterisation of the energy shift and its origin is essential for understanding of experiments on the asymmetric double well and is hence the main topic of the following sections.

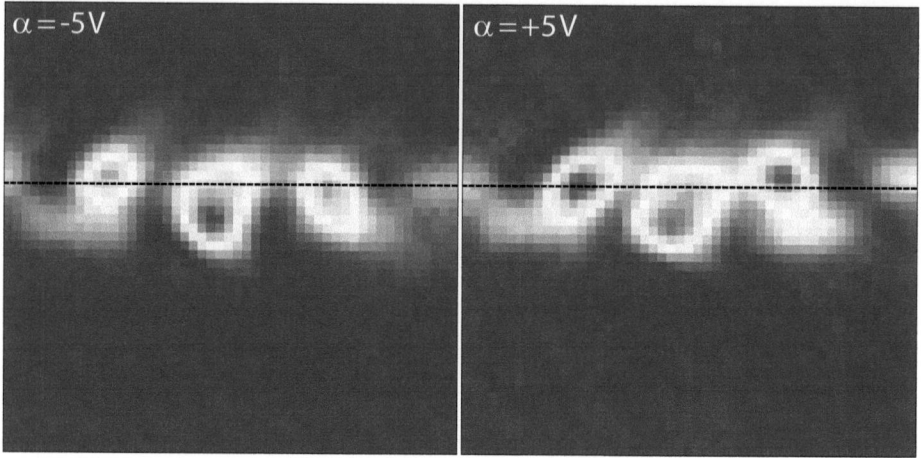

Figure 5.8: Averaged images of 10 experiments after 4 ms time-of-flight. The change in height of the two wells is clearly visible. The big cloud in the middle are atoms projected onto the $m_F = 0$ state which is insensitive to magnetic fields and keeps its position. Left: Asymmetry parameter $\alpha = -5$ V. Right: Asymmetry $\alpha = +5$ V.

5.4.2 Gravitational Potential Difference

Consider two atoms each with a mass m and in different sites of the double well potential. In the homogenous gravitational field g near the earth surface both have a certain potential energy depending on their position in height. To determine the gravitational energy difference in the double well potential we need to know their relative difference in height Δy. Since we are not able to measure Δy in-trap we calculate it from the splitting distance d and the tilt angle γ between the two clouds with respect to the horizontal. The gravitational potential then becomes

$$\Delta V_{\text{grav}}(\alpha) = mgd(\alpha)\sin[\gamma(\alpha)]. \tag{5.31}$$

Both the well separation d and the angle γ depend on the asymmetry parameter α. We derive the splitting distance and its behaviour with asymmetry in the last section. In order to get a complete picture of the asymmetric double well we discuss the determination of the tilt angle γ in the following. So far we described the asymmetry of the trapping potential with the artificial parameter α. The conversion to an angle, however, will give us a more concrete picture.

Tilt Angle

The tilt angle is difficult to access *in situ*, since the splitting distances are on the order of a few μm. Instead we make again use of the balancing scheme as described in section 5.3.3. After the rf turn off the atom clouds in the $m_F = 1$ and 2 states cross each other in the centre of the IP trap. The movement inside the static trap changes their relative distance while the relative orientation with respect to the camera chip is preserved. We image the clouds after a fixed time-of-flight of 2 ms for different asymmetries α and measure their relative tilt angle with respect to the horizontal which is defined by the camera chip. We achieve a better resolution by increasing the total rf field strength but keeping the ratio of the wire currents constant. The larger splitting inside the trap results in a higher kinetic energy gain while moving in the IP trap. Since the clouds move in opposite directions, they appear further separated while the tilt angle is maintained. Especially, in this measurement the dialed rf output powers are $-12.0\,\text{dBm}/-8.0\,\text{dBm}$ compared to the usual $-13.9\,\text{dBm}/-9.9\,\text{dBm}$ we use typically for interferometry. In figure 5.9(a) we present the measured tilt angle versus the asymmetry parameter. The fit indicates a slope of around $0.51 \pm 0.05°/\text{V}$. Especially, for large tilt angles the observed error is increased and the data points seem no longer to lie on a straight line. An additional fit through the central region yields a slope of $0.63 \pm 0.04°/\text{V}$ reproducing well the measured data for angles between $-4°$ and $2°$. However, an immediate conclusion on the correct tilt angle is impossible.

Influence of the Release

Considering the previous measurement the question arises if the angle changes during free fall as the release might not be smooth. In fact according to figure 3.7 the static wire current in the chip wires decreases to 0 A within $80\,\mu\text{s}$ which could cause a force onto the released atoms. A kick acting unequally on the wells influences the trajectory of the clouds and results in a change of the clouds' relative position. Especially at large splitting distances we have to take this effect into account.

A repetition of the previous measurement at various release times is therefore necessary. As shown in figure 5.9(b) the tilt angle changes indeed during free fall. Since $\tan\gamma \approx \gamma$ for γ in the order of a few degrees, the assumption that a small linear relative movement of the clouds implies a linearly changing tilt angle is justified. Hence the linear extrapolation of the data yields the initial in-trap angle. The extrapolation at $\alpha = 0\,\text{V}$ has with $0.33 \pm 0.05\,\text{rad/ms}$ a much smaller slope compared to that at the two extreme asymmetries of around $0.415 \pm 0.08\,\text{rad/ms}$. Fitting to the initial trap orientations shows a change in tilt angle of $0.60 \pm 0.06°/\text{V}$ which agrees well with the result for small angles in the previous measurement. It is also observable that for large tilt angles we find again a larger error. We conclude that the previously underestimated angle of $0.51°/\text{V}$ follows from the different behaviour after the release and the big uncertainty at the outer border of the measurement range.

Conversion

The determination of the tilt angle allows a direct conversion from the abstract and artificial parameter α to a more practical one. We obtain the in-trap orientation of the double well from the behaviour of the angle during free fall. Unfortunately, the limited amount of data in this set does not allow a meaningful derivation for the slope of the conversion. The observation that the tilt angle changes less after release for a small range of tilt angle, however, justifies partly the use of our first angle measurement where more data points link the connection between asymmetry parameter and tilt angle. Taking this considerations into account we find a conversion given by $\gamma(\alpha) = -1.05° + 0.63°\text{V}^{-1}\,\alpha$.
If we ask for the gravitational potential difference, however, we have to remember that the angle

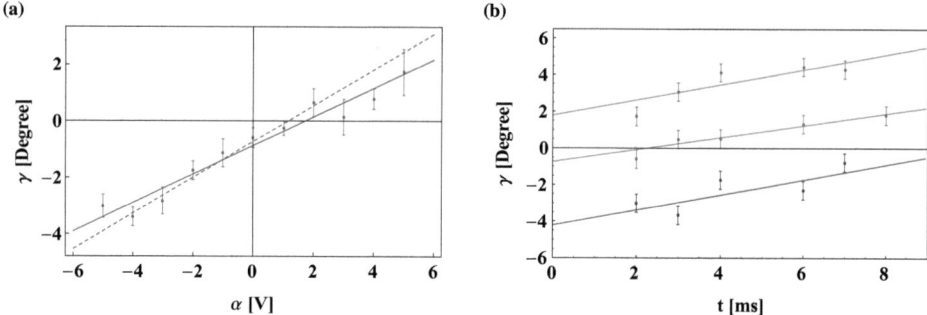

Figure 5.9: a) Measured angle versus asymmetry of the double well after 2 ms time-of-flight. The fit (solid line) shows a slope of $0.51°/V$ per asymmetry unit. A fit through the central data points (dotted line) reproduces the well the measurement. The angle seemed to be reduced especially for large asymmetries. b) An investigation of the angle for different time-of-flights shows that the angle changes linearly during free fall. The extrapolation to the initial tilt angle γ yields a slope of $0.60°/V$.

measurement is referenced with respect to the camera chip. The true orientation of the camera chip, direction of splitting and gravitational field with respect to each other is hardly accessible. Thus our angle conversion only delivers a relative gravitational shift ΔV_{grav} but not its exact absolute value. The point is that an angle of $0°$ in our conversion corresponds to $\Delta V_{\text{grav}} = 0$ according to equation (5.31) which might not be the case in reality because the orientation with respect to the gravitational field might be different.

5.4.3 Spectroscopy

Tilting the clouds out of the horizontal not only induces a gravitational energy shift, it also results in one cloud being closer to the chip surface than the other. Hence the cloud closer sees a larger rf dressing field b_{rf} and experiences a larger Rabi coupling Ω. We expect from calculations that this magnetic energy shift ΔE_{mag} is comparable in order of magnitude to the gravitational energy shift. The complete characterisation of the asymmetric double well requires an *in situ* measurement of ΔE_{mag} which we present in the following.

Experimental Procedure

We prepare the BEC in the double well potential as described in section 5.3.2 at an asymmetry $\alpha = 0$ V and a power ratio of $P_{Z4}/P_{Z3} = -13.9\,\text{dBm}/-9.9\,\text{dBm}$, the typical values for all interferometer experiments. After the preparation we tilt the potential to the asymmetry at which we want to probe the energy difference between the wells. We hold the atoms in this position for 30 ms while applying an additional rf field $\mathbf{b}_{\text{sp}}e^{i\omega_{\text{sp}}t}$. The field is generated by running a small rf current in Z configuration through the copper-H below the chip and is linearly polarized along the y axis. This is exactly the same configuration we also use for evaporative cooling. The rf field strength is with $|\mathbf{b}_{\text{sp}}| \approx 1.5 \times 10^{-4}\,|\mathbf{b}_{\text{rf}}|$

small compared to the dressing field which is necessary to prevent a shift in energy by deforming the adiabatic potential. The frequency of the spectroscopy signal defines a sphere on the double well potential where hot enough atoms are coupled out of the trap. If the probe field is resonant with the trap bottom of one of the potential wells, the atoms in the lowest energy level will cascade from their initial state to the untrapped states and will leave the confining potential. Monitoring the atom number with the frequency ω_{sp} of the probe field then gives us direct information about the level spacing of the dressed states. Since the five eigenenergies of the dressed potential are equally spaced, the magnetic energy on the resonance sphere is just twice the splitting $\hbar\omega_{sp}$ between adjacent levels.

We work with a red-detuned trap at fairly large rf coupling Ω in which case we expect that the transition rates for higher orders of n are more pronounced [106]. Instead of scanning the frequency over Ω_θ, we probe the transitions around the resonances for $n = 1$ with $\omega_{res} = \omega_{rf} + \Omega_\theta$ allowing high outcoupling at a small spectroscopy field strength. On the otherhand we have to avoid outcoupling of all atoms inside each well in order to get a sharp minimum in the spectral lines. Saturation of the transition would broaden the minimum and decrease the accuaracy of the measurement.

The spectroscopy on an asymmetric double well should show twice the amount of resonances of a single well trap. As the tilt angles are only of the order of a few degrees we expect that the resonances for the two wells are too close to each other to be distinguished. Therefore, we resolve the two sites after 2 ms time-of-flight after applying the balancing scheme at the end of the spectroscopic probing. The population in each well can then be counted independently for a certain rf frequency ω_{sp}.

Theoretical predictions for adiabatic potentials yield the transition frequencies at the minimum of the double well. In practice, however, the gravitational sag shifts the position of the atoms into an area of higher trapping potential. The displacement due to gravity is given by g/ω_r^2. Regarding that the double well potential has a modified radial trap frequency ω_r of 1.4 kHz [78] this displacement corresponds to 0.1 μm. Since the atoms move in an area of higher magnetic field gradient, the frequency range over which atoms are outcoupled is broadened. However, we do not expect that the trap bottoms of the two wells shift significantly relative to each other, since the displacement due to gravity is less than half of the characteristic radial extension of the wavefunction $a_r = (\hbar/m\omega_r)^{1/2}$ in one well which is around 0.3 μm.

Observed Transition Frequencies

We repeat the spectroscopy experiment at four different asymmetries $\alpha = -5$ V, -2 V, 0 V and 5 V. At each orientation we scan the frequency over the predicted transitions. The results of the four frequency scans are presented in figure 5.10 where we see typical resonance lines for each of the wells. We determine the minima in optical density from the Gaussian fits. At the minimum in optical density the spectroscopy field is resonant with the corresponding BEC and most of the atoms are outcoupled. The transition frequencies are nearly degenerate at $\alpha = -5$ V but show a clearly visible gap at $\alpha = 5$ V. We determine the energy shift for the different asymmetries and plot it against the tilt angle γ in figure 5.11 showing a linear increase. The magnetic energy shift has a slope of $\partial \Delta E_{mag}/\partial \gamma = 234.04 \pm 21$ Hz/°. The overall shift between $\alpha = -5$ V and 5 V is with around 1.4 kHz two times larger compared to the gravitational contribution with a total shift of around 0.7 kHz over the same range assuming a well separation of 3.23 μm.

Remarkable is also the variation in optical density at the population minima. The values of the minimum optical density are nearly equal at $\alpha = -5$ V. Increasing the asymmetry parameter the minimum of the left well increases while that one of the right hand side decreases. We explain this observation with a shift in coupling strength which is expected to be $\Omega_{sp}/2$ at the well centres. In the experiment, however, we have to consider the sag due to gravity. Depending on the orientation of the potential with respect to the horizontal the position of one cloud is shifted towards the IR while the other sags towards the OR. Since the coupling is antisymmetric around the well centre at $\theta = \pi/2$, the coupling frequencies are $\Omega_{sp}\left[1 - \text{sgn}(\Delta)\cos\theta\right]/2$ and $\Omega_{sp}\left[1 + \text{sgn}(\Delta)\cos\theta\right]/2$ respectively. In the resonance lines this asymmetry clearly manifests itself in the opposite outcoupling behaviour of the atoms measured for the two wells.

Furthermore, we observe an overall shift in transition frequencies which is due to the small shift in $|\mathbf{b}_{rf}|$ induced through the linear modulation of the rf sources with slightly different output power. We calculate the mean resonance frequency for each asymmetry and plot it against the tilt angle in figure 5.12. The fit indicates a slope of 2.64 ± 0.04 kHz/°. Remembering equation (5.22) we estimate an rf field strength of $|\mathbf{b}_{rf}| = 0.813 \pm 0.005$ G at the balanced position $\alpha = 0$ V or $\gamma = -1.05°$. Over a maximum tilt range from $-4.05°$ to $1.95°$ the field strength increases from around 0.77 G to 0.85 G showing a variation of around 5% in agreement with change variation we found in separation distance.

5.5 Conclusion

We implemented successfully rf adiabatic potentials in our atom chip trap. We demonstrated that it is possible to split a BEC equally into two wells. By adding/subtracting a small rf modulation to the rf currents in the chip wires we are able to tilt the double well potential with respect to the horizontal. The tilt introduces a small asymmetry between the wells in the form of a small potential difference. If the potential is rotated after the splitting, the size of the two clouds, however, stays constant. We identified the potential difference to be due to a height difference in the gravitational earth field and an increased (decreased) rf coupling of the atoms closer to (further away from) the atom chip surface.

In order to determine the gravitational contribution, a parametrisation of the asymmetric double well in terms of the tilt angle is necessary. The shift in magnetic coupling is gained from spectroscopy on the rf adiabatic potentials and is around twice the amount due to gravity. The analysis shows that the overall potential difference can be approximated by a linear function in tilt angle. This simple relation gives us easy control over the asymmetry of the double well potential.

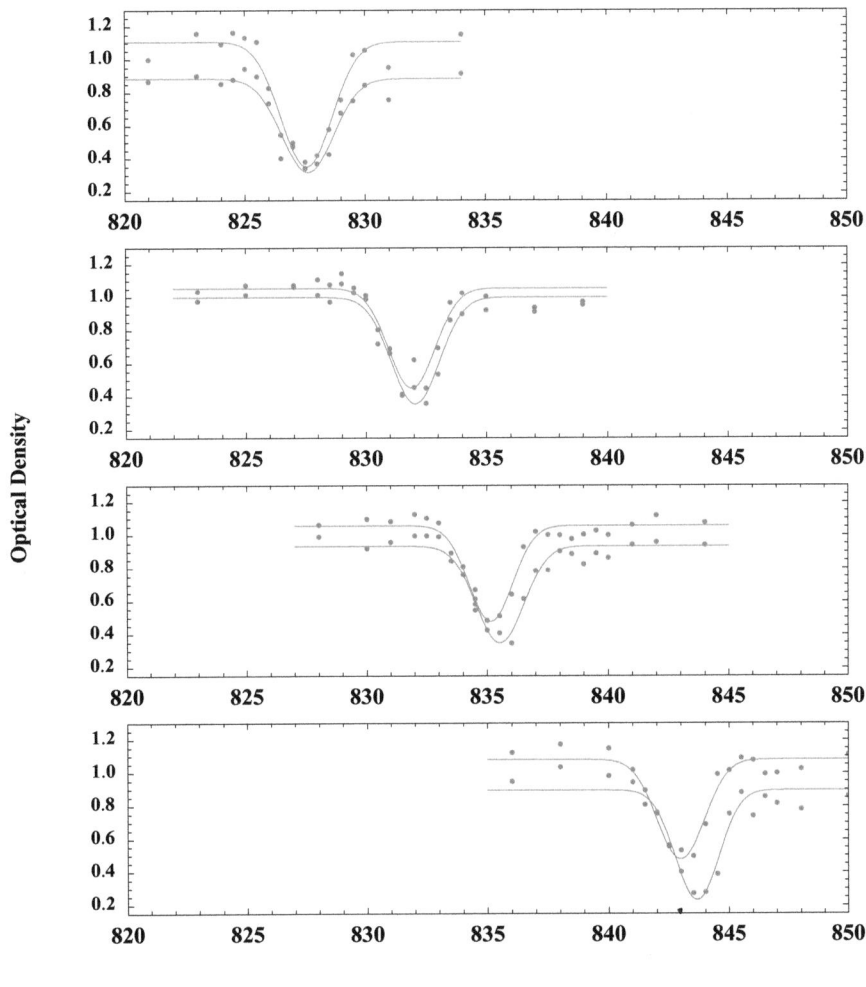

Figure 5.10: Typical resonance lines of the two wells at four different asymmetries $\alpha = -5\,\text{V}$, $-2\,\text{V}$, $0\,\text{V}$ and $5\,\text{V}$ from top to bottom. The frequency of maximum atom outcoupling is determined with gaussian fits. At $\alpha = -5\,\text{V}$ the transition frequencies of the left (dark grey) and right (light grey) well are nearly degenerated. An obvious gap is observed for larger asymmetry parameters. Due to a change in coupling strength the minimal value in population of the two sites behave oppositionally. Since with asymmetry we also make a small change to the total rf current, the mean transition frequency of the wells shifts over a range of around 15 kHz allowing an estimate of the rf field strength $|\mathbf{b}_{\text{rf}}|$.

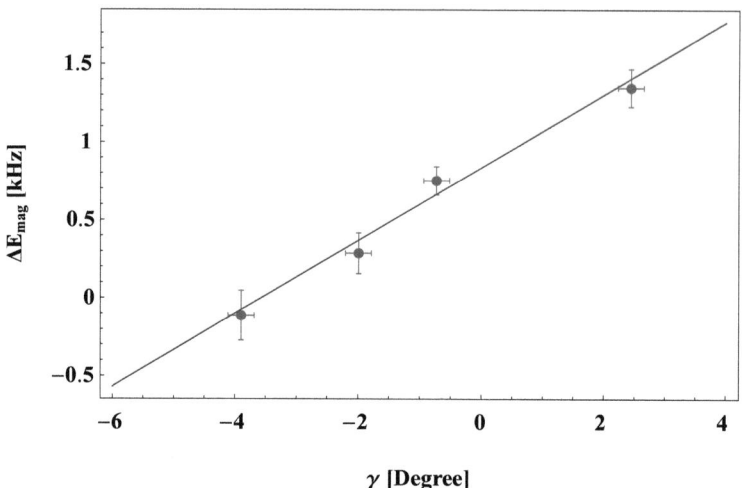

Figure 5.11: Magnetic energy difference between the left and right well plotted against the tilt angle γ. The overall shift of around 1.4 kHz corresponds to twice the amount of the gravitational potential gap for the implemented trapping configuration. The shift has a slope of 234.04 ± 21 Hz/° which is a crucial result for the characterisation of the double well potential.

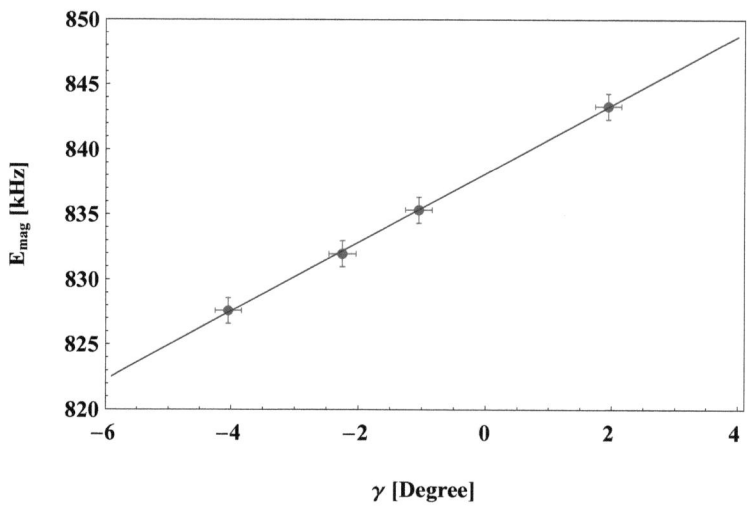

Figure 5.12: Shift of mean resonance frequency vs. tilt angle γ. The linear fit has a slope of 2.64 ± 0.04 kHz/°. The data allows an estimate of the rf field strength which varies in a range between 0.77 G to 0.85 G.

Chapter 6

Making the Interferometer Work

6.1 Introduction

Theory predicts the evolution of density modulations if two BECs are combined in free fall. The position of the modulations depends on their relative phase. The basic prerequisite for a reliable and repeatable phase measurement is that the splitting and recombination process of the BEC is coherent. This is not immediately clear, e.g. a very slow splitting in which the system has enough time to adapt itself to the changing environment would lead to a number state with a completely undefined phase. On the other hand any randomness like noise during the experimental sequence will also affect the outcome of the phase measurement. In the past a reproducible relative phase was reported in interference experiments of two BECs released from optical potentials [14] and from adiabatic potentials on an atom chip [13].

In this chapter we analyse the reliability and reproducibility of a phase measurement. We start with the observation of matter wave interference and how we determine the relative phase. In the next step we test if the splitting and recombination process is coherent. We discuss our studies and make improvements in the trap release which lead us to a more stable measurement. Finally, in section 6.5 we measure the dephasing rate between the two interferometer arms which is a crucial result for both the limit and the precision of our device.

6.2 Matter Wave Interference

The observation of matter wave interference is the most striking demonstration of the existence of a theory such as quantum mechanics. Its occurrance postulates both the wave nature as well as the phase property of a solid particle and cannot be explained within any classical theory. We consider one special case of matter wave interference occuring when two clouds of atoms all occupying the same ground state are overlapped and it was not until 1997 that this type of experiment was demonstrated [53] for the first time.

6.2.1 Observation of Interference

Interference Pattern

The emerging of the density modulations in the overlap of two BECs during free fall is presented in figure 6.1. We image the cloud on axis 2 along the z axis and the initial BEC is split as described in section 5.3 where the rf field is ramped up over 20 ms. We release the atoms from the double well potential by turning off the static magnetic field and the rf field at the same time. The density modulations within the cloud are clearly distinguishable after 12 ms free fall which is a typical time-of-flight for our experiments.

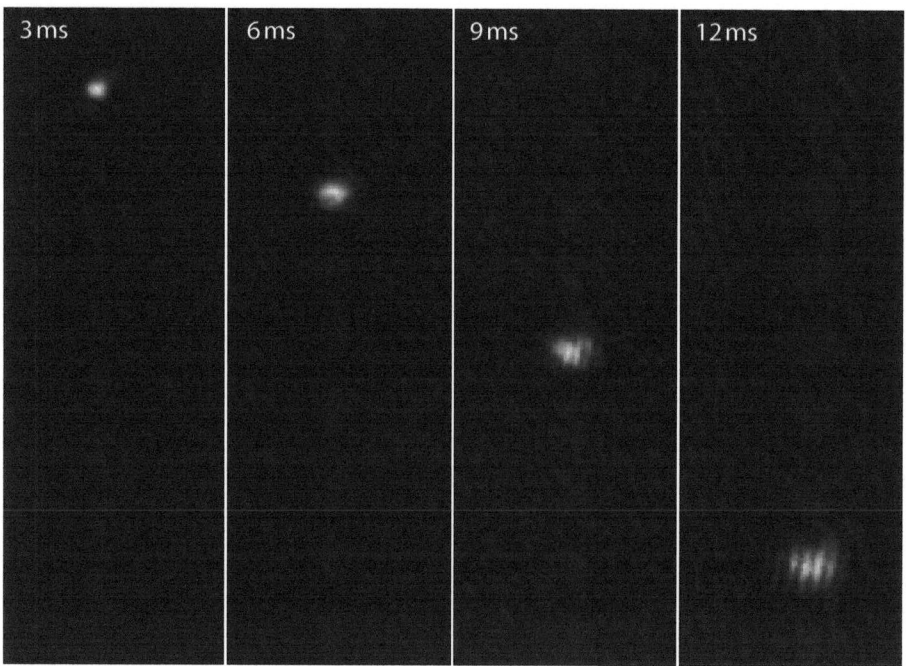

Figure 6.1: Time-of-flight images of two BECs released from a double well potential. The two clouds overlap in free fall and initially only a cloud with Gaussian profile is visible. First density modulations emerge after 9 ms time-of-flight and are clearly visible 12 ms after the release of the atoms.

Splitting Time

We linearly ramp up the rf field within a time of 20 ms. We choose this specific time from tests where we vary the splitting time for the interference experiment. At ramp times slower than 30 ms the contrast of the fringe pattern is strongly reduced. On the other hand the interference pattern seems to be distorted, if the BEC is divided faster than 5 ms.

In fact the shape and the length of the rf ramp influences the quantum state in which the two wells end up. In the theory part we discussed the binomial and number states. For splitting times $\tau \to \infty$ the system can adjust itself and ends up in a number state. In the case of a sudden divide it is impossible to tell in which well a single atom is located because the wavefunction of each atom is spread initially over the whole BEC and an adaptation is impossible. The result is a binomial distribution of the atoms. Between the binomial and number state other atom distributions are imaginable. Number squeezed states in multi well systems were realized by adjusting the splitting time of the BEC in the Oberthaler group [114]. Such states are of great interest for interferometry as they show increased coherence times [15].

In this thesis, however, we only consider one specific state of the atom distribution which is characterised in this chapter. However, to improve the precision of our interferometer in the future the study of different rf ramping shapes and times might be profitable.

Wavelength

We control the wavelength of the fringe pattern indirectly through the separation of the two wells. According to equation (5.26) a higher rf field strength leads to a larger splitting distance. At small enough rf field only one single fringe is observed, whereas a higher rf field decreases the wavelength as can be seen qualitatively in figure 6.2. The pattern with the large central peak defines the regime where the BECs are split into a double well potential but the chemical potential μ of the two BECs is still larger than the barrier height V_0 and both occupy the same wave function [115]. The relative phase is locked and an applied energy difference will result in a total and global phase shift. More fringes appear for $\mu < V_0$ and the tunnel coupling decreases proportional to $e^{-2\sqrt{2m(V_0-\mu)}d/\hbar}$. With the reduction of the tunnel coupling the BECs become independent and can acquire a shift in relative phase [116]. The limit for the minimal observed fringe wavelength is set by the resolution of the imaging system. Hence, at large separation lengths the density modulations appear no longer well defined.

In the past a discrepancy of equation (2.21) for small well separations was observed [13, 117, 118] which is explained with interaction effects [115] and the consideration of extended sources [78]. In terms of interferometry we are interested in measurements of the relative phase which in principle requires only that the two BECs are well separated and tunneling can be neglected. A proof that the wells are independent in the regime where we operate the interferometer is provided later in this chapter. However, in order to calculate the energy difference between the wells due to gravity we need to know their splitting distance. Since the interferometer requires independent BECs, we operate at fringe wavelengths much larger than that of the observed single fringe. For that reason we assume in the following that equation (2.21) is a good approximation.

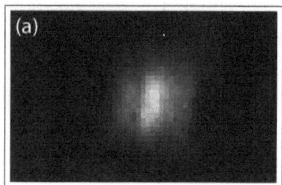 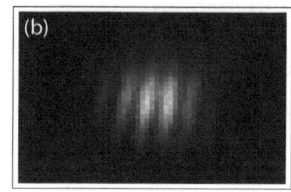 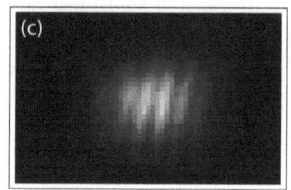

Figure 6.2: Interference patterns after 12 ms time-of-flight for three different rf field strengths. (a) The single fringe occurs at barrier heights where the wavefunctions of the BECs still overlap in the potential. At larger separation length more fringes appear whose wavelength decreases with larger rf fields. The fringe pattern in (b) is typical for the field strength at which we operate the interferometer and shows a well defined fringe pattern. At smaller wavelengths (c) the resolution of the camera starts limiting the visibility of the density modulations.

6.2.2 Extracting Information

In order to extract information from images of the density modulation we cut the image perpendicular to the fringes. The cut is integrated over several pixels parallel to the fringes which smoothens local fluctuations in density. To the resulting one dimensional array we can then fit the function

$$F(x) = A \exp\left[-\frac{(x-x_0)^2}{r^2}\right]\left[1 + v\cos\left(2\pi\frac{x-x_0}{\Lambda} + \phi\right)\right] \tag{6.1}$$

which satisfies the physics of equation (2.20). The first, Gaussian, part of the function describes the total overlap of the two BECs where x_0 is the centre of the cloud and r its width. The second part describes the density modulations with wavelength Λ and the contrast v of the pattern with respect to the maximum A of the Gaussian function. The contrast can be reduced by the integration [119] if the fringes are not perfectly perpendicular to the cut at any point. Typically, we integrate over only 5 pixels of the central region as marked in figure 6.3 in order to prevent such effects.

The measurement of the relative phase ϕ depends on the reference point of the fit. A different reference point will deliver a different releative phase. The true relative phase, however, can only be determined if the position of the fringes is compared with the centre of the double well potential. Since the position of the trap centre is not accessible, we have to choose a point which allows a stable phase measurement under shifts in the experiment, i.e. movements of the imaging beam or the camera. Therefore we reference the relative phase to the cloud centre x_0. The phase measurement is then stable under a spatial transformation of the interference pattern.

Since the phase obtained from the fitting process is of circular nature, we have to use the mathematical tools provided by circular statistics to calculate the mean value, the standard deviation and the variance in repeated experiments [120]. A short overview of the methods used in this thesis is given in the appendix.

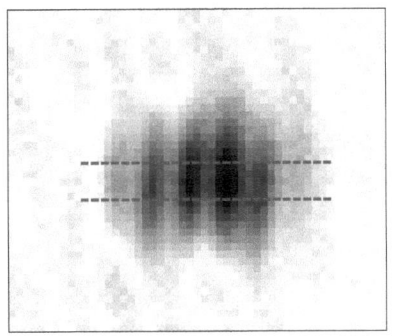
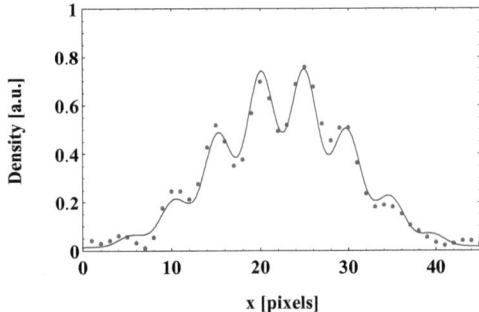

Figure 6.3: For a quantitative analysis of the interference pattern we cut the interference pattern perpendicular to the fringes and integrate over the central region marked by the dashed lines. The result is a one dimensional array of data points to which we apply an appropriate fit.

6.3 Spin States

Before we analyse the coherence of the interferometer we spend some thoughts on the release from the adiabatic potential. The dressed atom states are a superposition of the bare magnetic spin states. At the point in time when the rf field vanishes the dressed atoms are projected onto their bare magnetic spin states. Although initially the atoms are prepared in the $m_F = 2$ state, this might be no longer true after the projection. Since the current in the chip wires does not vanish immediately after its control voltage is switched to 0 V, there are magnetic fields near the chip surface lasting for several μs which can affect the release of the atoms. We mentioned already in section 3.2.6 that to guarantee the overlap of the two wells a precise balancing between turn off time and offset current in the chip wires is necessary. Since different spin states experience a different force in a magnetic field, the goal is to project all atoms onto the same state. A variable force over the two atom clouds not only influences their overlap but also the phase of the atoms. Both is undesireable for an atom BEC interferometer.

6.3.1 Projection

The composition of spin states depends on the angle

$$\theta = \arctan\left(-\frac{\Omega}{\delta}\right) \tag{6.2}$$

which the effective magnetic field encloses with the quantization field. Maximal projection onto the $m_F = -2$ state is realized for angles $\theta = \pi/2$ orthogonal to the original state which happens for rf frequencies resonant with the trap bottom. In the case of a decreasing static magnetic field the absolute value of δ becomes large with respect to Ω and equation (6.2) tells us that $\theta \to 0$. For $\theta = 0$ we expect a projection onto the initial state. If the detuning δ becomes large, however, the rotating wave approximation breaks down and the simple equations are no longer applicable for a quantitative analysis.

Studies on the spin projection in adiabatic potentials in connection with the ramp down time of the rf field were made by J. J. P. van Es [118]. However, no complete picture can be found in the literature. In the past also the role of the phase of the rf field was discussed. We examine the influence of the relative turn off time between static and rf fields, and hence the detuning δ in the following.

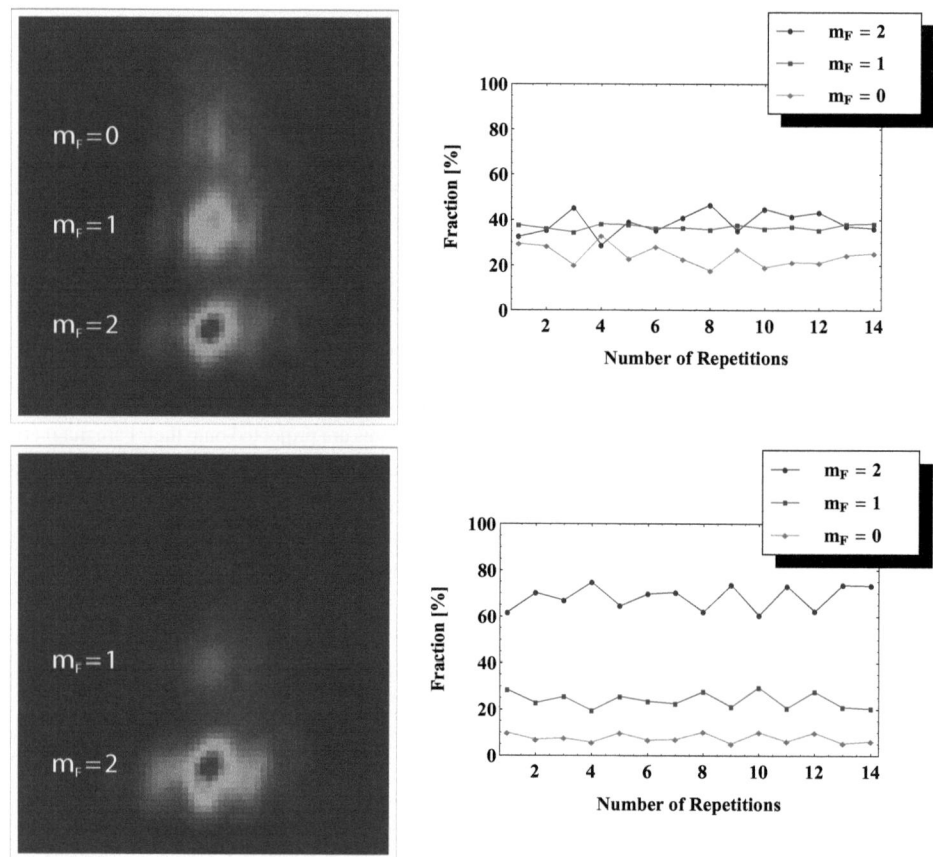

Figure 6.4: Stern-Gerlach experiment on the overlapping double well clouds. When the rf field is turned off, the atoms are projected onto their bare magnetic states. The distribution of the projection depends on the detuning δ. The single shots show that the process is reproducible. Top: Without the rf delay atoms we observe atoms in the $m_F = 0, 1$ and 2 state. Bottom: Implementing the rf delay leads to an occupation of around 70% in the $m_F = 2$. No atoms occupy the $m_F = 0$ state.

6.3.2 Observing Spin States

In order to measure the spin state composition we perform the interference experiment as described in section 6.2.1. We implement the double well at a splitting distance where the wave functions still overlap and the interference pattern shows only one pronounced peak. During the free fall of the atoms we now apply a magnetic field gradient along the y axis which separates the different spin states during time-of-flight analogue to the Stern-Gerlach experiment.

The rf current is controlled by a switch with a response time faster than $1\,\mu s$. The fast switch effects the rf field to vanish immediately while the chip wire field drops within $80\,\mu s$. The projection happens in the environment of the static magnetic trap with $\delta = 90\,\mathrm{kHz}$.

For comparison we take a second data set using another turn off scheme. We keep the atoms dressed until the static magnetic fields have completely vanished. The properties of the control software set a limit to the minimum delay time. The rf switch is therefore triggered $400\,\mu s$ after the chip wire turn off yielding a projection of the atomic spin states at a large detuning δ.

In figure 6.4 we show the averaged images each from 14 repeated experiments. If we turn off the trap at small detuning δ we can see atoms in three different clouds. The weakest of them is located at the position of the original interference pattern marking the $m_F = 0$ state. We find 37% of the atoms in the $m_F = 1$ state and a pronounced $m_F = 2$ cloud. Spin states with negative sign are not observed, as they are immediately expelled from the trap during the turn off. The picture is quite different if the rf delay is implemented. In this case we find no atoms in the $m_F = 0$ state and a very weak cloud for $m_F = 1$. Most of the atoms, around 70%, are projected onto the $m_F = 2$ state. We also plot the fraction of the spin states found in the single shots of each experiment. The distribution is stable and shows small fluctuations around the averaged values leading to the conclusion that the projection is repeatable and independent of the phase of the rf current.

6.4 Phase Coherence and Stability

Two coherent BECs have a non-random relative phase. If we prepare two BECs independently from each other we end up in a so-called Fock state. The number of atoms in each cloud is then well defined, but their relative phase is completely random. Since an initial phase is not defined, it is impossible to extract an applied phase shift from an interference pattern. Therefore the preparation of coherent BECs is an absolutely necessary requirement for an atom BEC interferometer.

In this section we study the repeatability of the fringe pattern in our experiment. We show that the splitting process is coherent and repeatable. We find that the width of the phase distribution is also governed by the turn off of the adiabatic trap and a comparison between the two turn off schemes is presented. Furthermore, it turns out that a correction can be applied to achieve small phase spreads and precise phase measurements.

6.4.1 Phase Distribution

In order to test if the BEC is split coherently, we ramp up the rf field over 20 ms and release the atoms immediately, without any hold time after the splitting process has finished. We take an absorption image after 12 ms time-of-flight and extract numbers by fitting the function of equation (6.1) to the density modulations. The experiment is repeated 100 times which allows us not only to compare the relative phase but also to check the stability of fringe spacing, visibility, centre position, etc..

Synchronized RF Turn Off

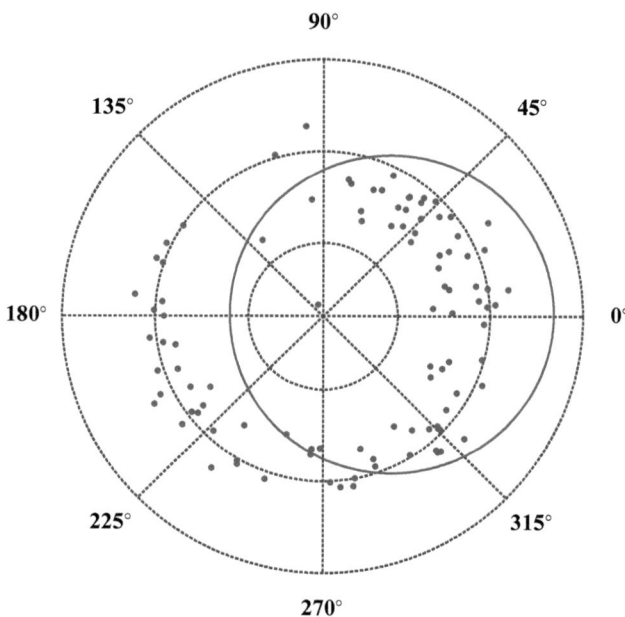

Figure 6.5: Phase distribution of 100 repeated experiments with fast rf turn off. The radius of the data points represents the fringe visibility. The contrast is with $v = 0.21 \pm 0.03$ stable. With a circular standard deviation of around 100° the data points are spread over the whole circle. The von Mises distribution (solid line) is fitted to the counts within small sections of the circle. It shows qualitatively a characteristic direction representing the mean of the phase.

In a first test of phase coherence we switch off the rf field and the magnetic fields of the IP trap simultanously. The dressing field vanishes much faster than the static magnetic fields and we project onto at least three different spin states. Figure 6.5 shows a broad phase distribution with a circular standard deviation of around $\sigma_\phi = 100°$ or 1.75 rad. The radial component of the polar plot represents the contrast v which has a mean value of $\bar{v} = 0.21 \pm 0.03$. The stability of the fringe pattern concerning the centre positon of the cloud x_0 and the wavelength Λ is presented in figure 6.6. The wavelength is, with the exception of small fluctuations, constant over the whole data set and has an average of

5.15 ± 0.1 pixel corresponding to 18.84 ± 0.3 μm. The experiment runs with a constant periodicity

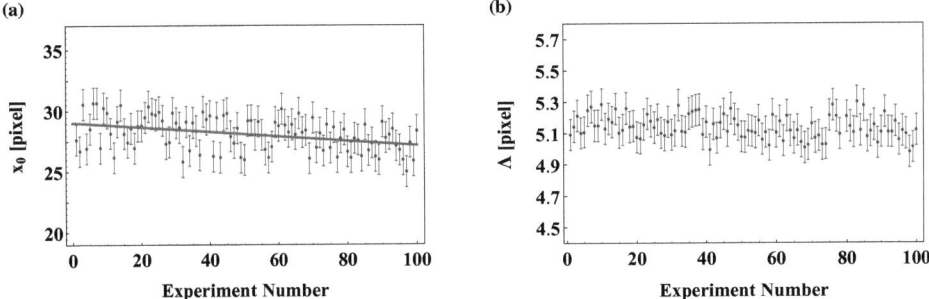

Figure 6.6: (a) The linear fit to the cloud centre positions x_0 has a slope of 17.3×10^{-3} pixel per experiment. The cloud centres x_0 are distributed around the fit with a standard deviation of 1.27 pixel. (b) The behaviour of the fringe wavelength Λ is stable from shot to shot and fluctuates around a mean value of 5.15 ± 0.1 pixel. A constant fringe spacing implies that the splitting distance is repeatable.

taking one data point every two minutes. In the data set the centre position x_0 of the Gaussian overlap clearly shifts with a linear drift of 17.3×10^{-3} pixels per experimental cycle adding to a total drift of 6.33 μm after 100 repetitions. The linear fit in figure 6.6(a) represents the average shift in time around which the measured centres are distributed.

The overall drift in centre position suggests that either the imaging beam is slowly drifting over the course of the experiment, or that the drift of the cloud centre is correlated with a change in the position of the double well potential. First has no effect on the phase measurement as the fringe fit is referenced to its Gaussian centre, and second is very unlikely as the resulting drift would require a change in the lab field of around 1 G which is larger than everything we measure near the experiment. The phase, however, is affected by variations in the turn off of the trapping fields. A random kick during the release influences both the relative phase and the centre position of the interference pattern.

Keeping this considerations in mind we calculate the expected cloud position for each experimental cycle from the linear drift and subtract the result from the actual centre position x_0. By doing so we obtain a set of the new positions x_0' distributed around the new mean position $\bar{x}_0' = 0$ μm as presented in figure 6.10(a). Finally, we plot the measured phases against the corrected cloud centres in figure 6.7(a). The Pearson's correlation coefficient of the data has a value of 0.85 suggesting strong correlation between the two parameters space and phase. The linear fit yields a slope of around 1.2 rad/pixel which allows a correction of the phase information for its deviation. This procedure delivers the corrected distribution in phase space shown in figure 6.7(b). The von Mises distribution fitted to the new phase distribution yields a circular standard deviation of around $\sigma_\phi = 12°$ which is similar to the results of T. Schumm et al. [13] with an analogous setup. We conclude that the splitting process is coherent. The correction and our considerations, however, suggest that the release from the trap has a random component with a strong impact on the relative phase. Still, a stable phase measurement can be realised.

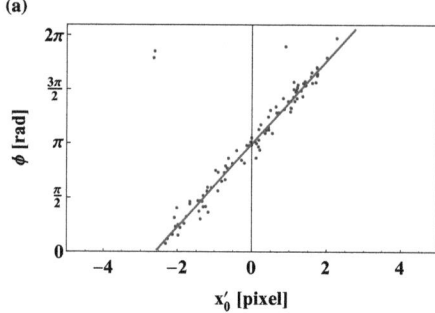 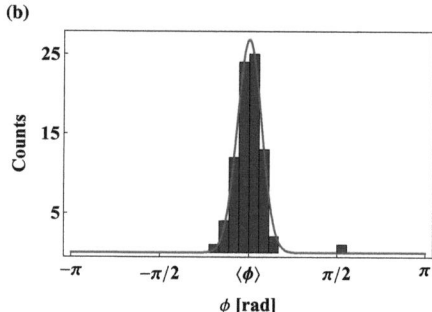

Figure 6.7: (a) Correlation between cloud centre position and relative phase. The relative phase is spread over the whole range of 2π because the cloud centre has a large variation from shot to shot. Fitting to the data points, however, yields a function which can be used to correct the phase in dependence of the cloud position in space. (b) The corrected phase distribution is clearly non-random and has a circular standard deviation of only around 12°.

Delayed RF Turn Off

Although the correction delivers a coherent phase measurement, it is not convenient in an experiment where parameters are varied. The problem is caused by the overall linear drift of the cloud position on the image. We found that its slope can change across a large data set and differs from day to day. Another requirement is that the experiment has to run periodically without longer breaks which is also not guaranteed due to typical problems such as the loss of the laser lock. In fact it is more practical to operate the interferometer in a mode delivering a stable and coherent phase measurement directly from the fits to the cloud. Since we concluded that the correlation between spatial position and relative phase is due to the turn off of the trap, it stands to reason to compare the previous results with data taken where the rf switch is triggered with a delay of 400 μs with respect to the DC current turn off. One should notice that the time-of-flight in this experiment is increased by the delay to 12.4 ms.

The analysis of 100 repetitions shows a circular standard deviation of the phase of around $\sigma_\phi = 48°$ or 0.83 rad which is half of the width in the previous experiment. The wavelength and centre position data sets are presented in figure 6.8. Visibility and fringe spacing are with mean values $\bar{\nu} = 0.12 \pm 0.03$ and $\bar{\Lambda} = 4.81 \pm 0.1$ pixel or 17.6 ± 0.3 μm again stable during the experiment. As expected the drift of the cloud centre is still observed and has a slope of 12.2×10^{-3} pixel per experimental cycle. Correcting the cloud centres for the drift and determining their new positions x'_0 we find that the distribution is much smaller than without the delay. The corrected phase distribution, however, has a width of around $\sigma_\phi = 23°$.

Finally, with an uncorrected $\sigma_\phi < 1$ rad we found a regime where a non-random relative phase can be extracted directly from the fit without applying any corrections. Hence the splitting process is not only coherent, moreover it is also possible to determine a certain relative phase from a small number of repetitions.

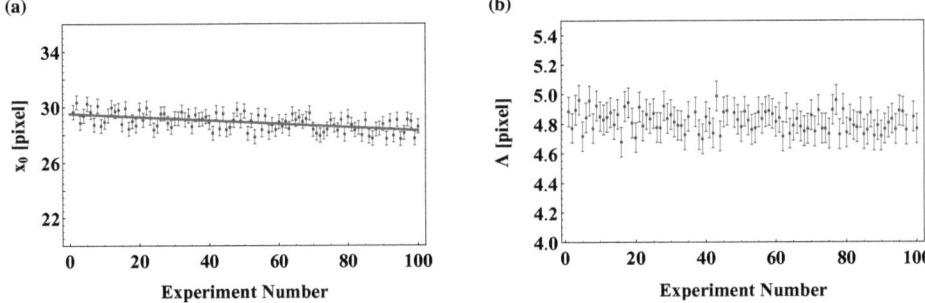

Figure 6.8: (a) In the case of a delayed rf turn off the cloud centre shows also a linear drift. However, the distribution of the spatial positions around the fit is much narrower. (b) The wavelength has a mean value of 4.8 ± 0.1 pixel and is stable over the course of the experiment. This is larger than in the previous data set due to higher rf power on the oscillator outputs.

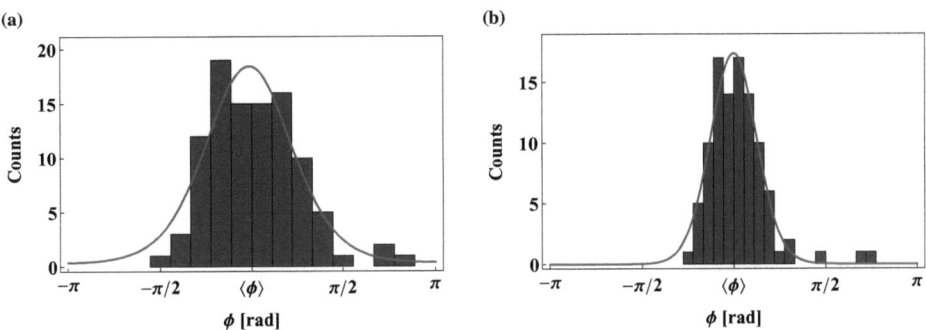

Figure 6.9: (a) The circular standard deviation is 48° with a delayed rf turn off. The phase is clearly not randomly distributed and its mean can be determined from the fit data. (b) The correction using the correlation between phase and space reduces the phase spread further to 23°. The corrected data represents the distribution of the physical state of the system.

6.4.2 Comparison

In the previous experiments we found that the visibility and the fringe wavelength are stable in repeated experiments. Especially, a constant wavelength is important as it is a direct measure for the cloud separation in the double well trap. Reproducibility of the trapping potential is therefore guaranteed.

The circular standard deviation of the corrected phase distribution differs by around 10° in the two tested schemes. This can be understood by the larger well separations in the experiment [13] with the delayed rf turn off. Since the splitting time is equal in both data sets, the two wells lose their connection at different points in time during an experimental cycle and the phase distribution starts spreading earlier. The analysis of the phase spreading is discussed in the next section.

In order to explain the discrepancy in the uncorrected phase distribution we have a closer look at cloud centre positions x'_0 corrected for the total linear drift. The standard deviation of the corrected cloud centre is around 4.3 pixel without the delay and only 1.8 pixel in the second case. The direct comparison is presented in figure 6.10. It is remarkable that we find an equal correlation between phase and centre position in both cases. It turns out that a smaller distribution in space results in a smaller distribution in phase. The fit yields a slope of 1.20 ± 0.04 rad/pixel which corresponds to a phase shift of nearly 2π over one wavelength. The origin of such a correlation is that while the Gaussian overlap of the interference pattern is moving from shot to shot the position of the fringes stays constant. Still the relative phase seems to change because the measurement is referenced to the cloud centre. Since no additional phase shift seems to appear, we conclude that the turn off kick must be acting equally on both wells.

The correlation between phase and space not only broadens the phase uncertainty σ_ϕ but also sets a lower limit to its width. The quantum mechanical uncertainty of the physical state $\Psi(\mathbf{r}, t)$, however, has to be determined from the described correction. Switching off the static magnetic fields has a

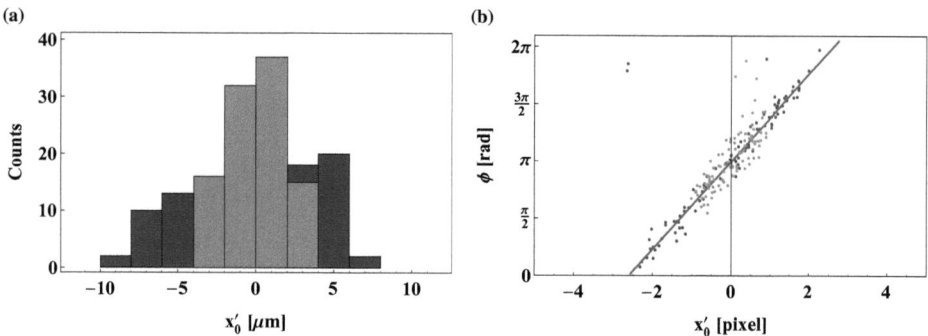

Figure 6.10: (a) The distribution in space depends on the turn off of the magnetic fields. Keeping the atoms dressed until the static magnetic fields have vanished highly decreases the variation of the cloud centre in the repeated experiment. Dark grey: rf field is turned off at the same time as the IP trap. Light grey: rf field is turned off $400\,\mu$s after the IP trap. (b) In both cases we find the same relation between phase and space. A smaller distribution of the cloud centre positions yields a smaller phase distribution. This correlation limits the phase spread to a minimum width, although the phase spread of the physical state might be much smaller.

random component affecting the trapped atoms. Projecting the atoms onto a single magnetic spin state delivers a more stable interference pattern. The computer control software limits the delay to a minimum of $400\,\mu$s. Additional tests could not prove a difference for larger delays in agreement to the expectation that the static magnetic fields vanishes completely within $100\,\mu$s. In all following experiments we therefore realise the trap release with an rf switch delay of $400\,\mu$s with respect to the turn off trigger of the IP trap and observe all interference patterns after $12.4\,$ms time-of-flight.

6.5 Limits of the System

So far we have shown that the splitting and recombination processes are coherent. However, in order to apply a phase shift between the two interferometer arms some interaction time between splitting the BEC and releasing the clouds from the wells is necessary. Equation (2.5) tells us that the resulting relative phase is the integral of the potential difference between the wells over time. In the case of tiny forces acting on the wells the sensitivity to measure such a force is hence limited by the interaction time. An infinite long movement of the atoms in the interferometer arms would mean that we are able to measure an infinite small potential difference. Of course, this is not feasible in practice, since the apparatus is influenced by quantum physical properties of the atoms and various external factors.

In general there are two effects which limit the maximum interaction time, phase spreading and a loss in fringe contrast due to phase fluctuations along the length of the clouds. Especially, phase fluctuations are enhanced by the tight confinement and the elongated geometry. We have already discussed the basic theory of these effects. In the following we turn ourselves to the experimental analysis and characterise the limit of our apparatus.

6.5.1 Phase Spreading

In order to measure the phase spreading we determine the distribution of the phase similiar to the coherence test in section 6.4 in repeated experiments. After the splitting process has finished, however, we now hold the BECs inside the double well for a certain time before turning off the trap. We take data sets for different hold times starting with $t = 0$ ms defined by the end of the splitting ramp.

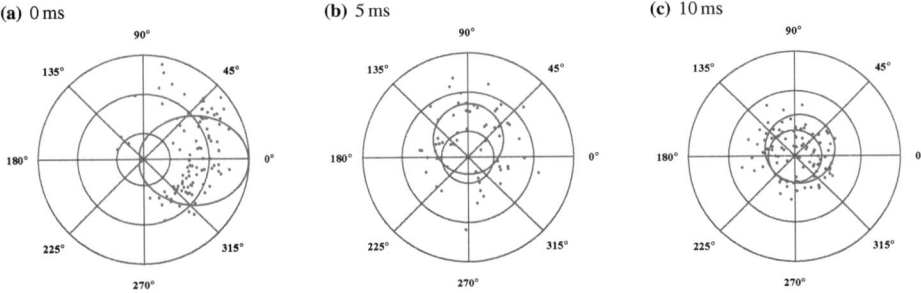

Figure 6.11: Polar plots for the relative phase. The different hold times inside the double well are from left to right 0 ms, 5 ms and 10 ms. Along the radial direction we plot the contrast of the interference pattern. The von Mises distribution visualises the phase spread. Its radial direction is in arbitrary units. Initially we find high visibility with a distinctive direction of the phase. After 10 ms we find that the von Mises distribution is nearly concentric with the coordinate circles of the plot suggesting a random phase spread.

To begin with we take 100 data points at 0 ms and 10 ms each, and another 61 repetitions at 5 ms interaction time. The phase distributions at these points are visualised in figure 6.11 where the contrast is again represented by the radial component. The von Mises distribution gives an idea of the randomness of the data. The initial, distinctive direction of the phase is completely lost after 10 ms and the mean fringe contrast drops to around 5.4%. After 5 ms the circular standard deviation of the phase distribution is already on the order of $\pi/2$.

Since we are interested in the behaviour of the physical state and not in technical noise limiting the distribution to a minimum width of 48°, we apply the discussed correction of the systematic error to the data. Additionally, we take 20 repeated experiments at another three hold times.

We plot the circular standard deviation σ_ϕ of the relative phase against hold time in figure 6.12. We find a linear growth of $\sigma_\phi = \frac{1}{\hbar}\sigma_{\Delta\mu} t$ according to equation (2.42) which corresponds to an uncertainty in the relative chemical potential $\sigma_{\Delta\mu}$. The linear fit yields an evolution of the phase spread with a rate of 161 ± 18 rad/s or an uncertainty in chemical potential difference of $\frac{1}{\hbar}\sigma_{\Delta\mu} = 26 \pm 3$ Hz. The initial phase spread at the end of the splitting process is $\sigma_{\phi,0} \approx 0.4$ rad and the dephasing criterion $\sigma_\phi(t) = 1$ rad is fulfilled after around 4 ms.

Extrapolating the fit back in time shows that it intersects the time axis at 2.4 ± 0.1 ms before the splitting process has finished. We estimated before that the double well forms within the last 5 ms of the rf ramp. Since initially the phase distribution of the wells is still locked by tunnel coupling, the extrapolation seems a reasonable estimate of the time at which the wells decouple.

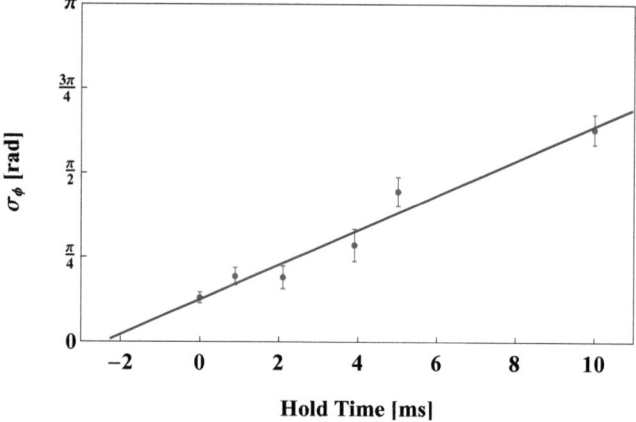

Figure 6.12: The phase spread measured against hold time. Each of the data points represents 20 repetitions except for the sets at 0 ms and 10 ms where we take 100 data points, and 61 data points at 5 ms. We find a linear growth of the circular standard deviation. The solid line is a fit with a slope of 161 ± 18 rad/s.

In figure 6.13 we calculate the averaged images for the three data sets with a large number of repetitions. The visibility decreases with hold time as expected from equation (2.38). Before averaging the single images were corrected for the linear drift of the cloud centre. Initially the contrast is around 11.4% which is close to the mean visibility of the single shots of 11.8 ± 0.1%, whereas it decreases to 3.7% after 5 ms hold time and 0.6% after 10 ms. From equation (2.38) we determine the expression for the phase spread according to

$$\sigma_\phi = \sqrt{-2\ln\left(\frac{v_{av}}{v_{single}}\right)} \qquad (6.3)$$

where v_{av} is the is the visibility of the averaged cloud and v_{single} the mean single shot contrast. Considering the drop in contrast the calculated phase spreads are 15°, 72° and 120° in good agreement with the corrected circular standard devitation at hold times of 0 ms, 5 ms and 10 ms as expected.

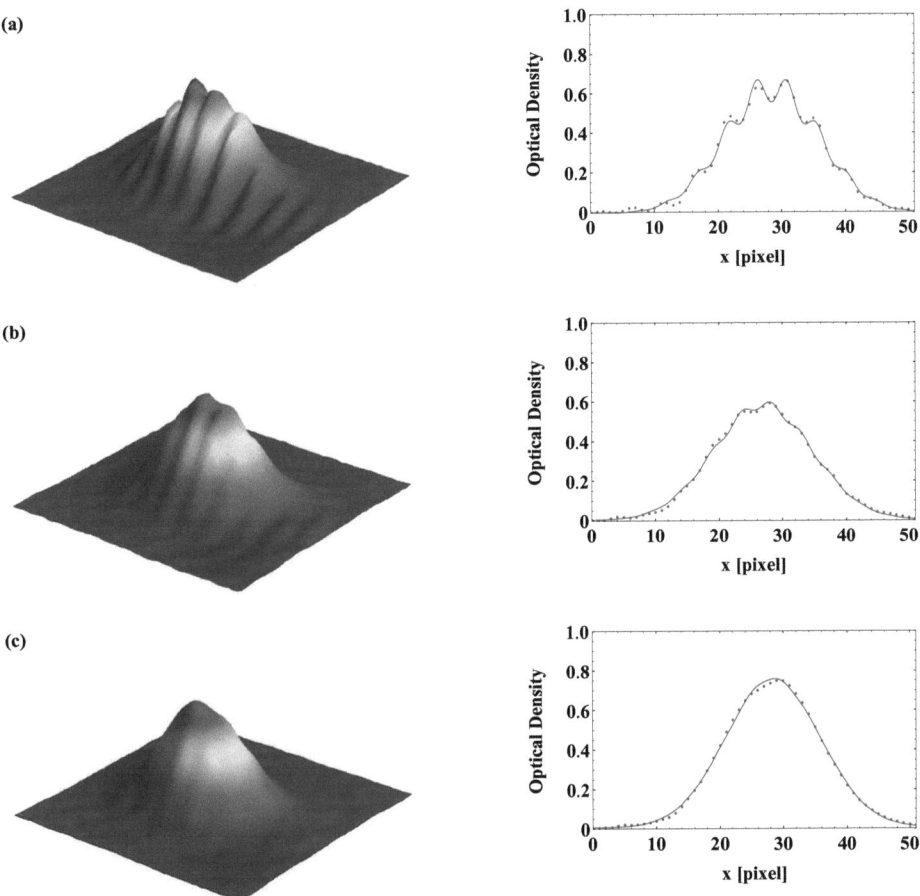

Figure 6.13: Averaged images of 100 (61 at 5 ms) equal experiments for three different hold times inside the double well potential. The visibility for (a) 0 ms hold time is with 11.4% nearly equal to the single shot visibility. It strongly decreases after (b) 5 ms. After 10 ms hold time the phase is completely random and the visibilty of the ensemble average drops below 1%.

6.5.2 Fringe Contrast

Besides a random phase distribution another factor that prevents a precise measurement is the loss in contrast of the fringe pattern. As soon as the contrast drops too far, fitting to the fringe pattern will no longer work reliably. Indeed we observed in the last experiments a drop in visibility from initially 11.4% down to 5.4% if the clouds linger 10 ms inside the potential. A loss in visibility can be caused by the two effects we discuss in the following.

Excitation of Dipole Modes

We split the BEC with a rate of 50 Hz whereas the longitudinal trap frequency of the magnetic trap drops from 28 Hz to around 18 Hz in the double well potential. Although the rf ramp is not resonant with the longitudinal trap frequency, the excitation of longitudinal oscillations is possible. A relative motion between the two clouds can lead to a twisting of the resulting interference pattern as shown by simulations [121]. Since we work with an elongated cloud, a variation in the twist along its length would yield a reduced contrast. Indeed an analysis of the movements in the double well potential revealed the existence of dipole and breathing modes and is therefore an issue.

Longitudinal Phase Fluctuations

Another reason for the interference pattern to blur are phase fluctuations causing the BEC to break up into domains of local phase. Immediately after the splitting process we find two identical copies of the BEC. Although the phase is not constant within one single cloud, their relative phase is equal along the length of the trap. Since the phase distribution in each quasicondensate is narrow, the fringe contrast is not affected. Phase spreading, however, enhances the discrepancy of phase measurements between the local domains along the length of the cloud. Absorption imaging along the z axis integrates over the different results and decreases therefore the contrast.

A direct observation of longitudinal phase fluctuations for two interfering BECs is not possible in our setup, but the assumption of its existence can be justified by an aspect ratio of ≈ 77 [62]. The number of local domains $\xi = T/T_\phi$ in which a BEC decays depends on its actual temperature T. At small chemical potentials the temperature is difficult to determine due to a lack of thermal background atoms. Hence, we use a different analysis.

The Thomas-Fermi approximation delivers equation (2.44) for the uncertainty in relative chemical potential $\sigma_{\Delta\mu} = \hbar \left(\frac{72}{125}\frac{m}{\hbar}\right)^{1/5} \omega^{6/5} a_s^{2/5} N^{-1/10}$ where a_s is the s-wave scattering length. Assuming a total number of atoms $N = 15 \times 10^3$ and a geometric mean frequency $\omega = 2\pi \times 360$ Hz we calculate a dephasing rate of $\frac{1}{\hbar}\sigma_{\Delta\mu} = 19$ Hz which is clearly smaller than the measured rate. The Thomas-Fermi theory also yields the relation $\sigma_{\Delta\mu} \propto \mu/\sqrt{N}$. But in the vicinity of phase fluctuations N becomes the number of atoms in the local phase domains while the chemical potential μ remains unaltered. Hence the dephasing time is enhanced due to phase fluctuations by a factor of $\sqrt{\xi}$. In our case we find a ratio between experiment and theory of around 1.4 suggesting that the BEC is divided into $\xi = 2$ local phase domains.

Phase Uncertainty vs. Contrast

After identifying the existence of longitudinal phase fluctuations we now study the phase uncertainty and the fringe contrast in dependence of the atom number (or chemical potential). Therefore we vary the evaporation stop frequency and release the atoms from the double well without any additional hold time.

The evaporation stop frequency influences not only the atom number but also the temperature of the BEC and the chemical potential which behaves accordingly to $\mu \propto N^{2/5}$. These parameters influence the occurance of longitudinal phase fluctuations. We expect to improve both phase spread and fringe contrast by adjusting our system.

We monitor the circular standard deviation of the phase and the fringe contrast as presented in figure 6.14. The phase uncertainty is smallest for chemical potentials larger than 8 kHz and increases clearly for smaller number of atoms. At a chemical potential of around 2 kHz the width of the distribution nearly doubled. This behaviour agrees with the observations of G.-B. Jo et al. [23] who proved that interferometry is still possible in the vicinity of longitudinal phase fluctuations. Without longitudinal phase fluctuations we expect that $\sigma_\phi \propto \mu^{-1/4}$, a behaviour which is visualised by the fit in figure 6.14(a).

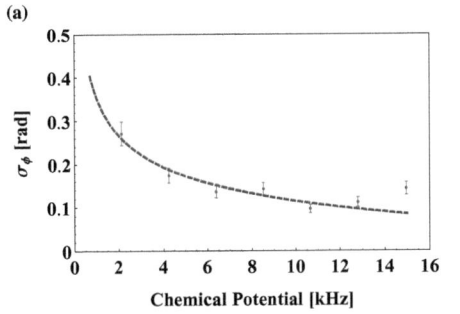
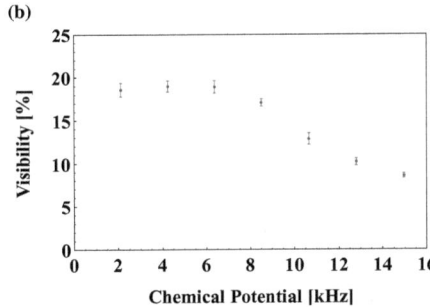

Figure 6.14: Behaviour of the interference pattern with chemical potential. Each data point represents 7 repetitions. (a) Circular standard deviation of the phase plotted against the chemical potential. For decreasing chemical potential the circular standard deviation of the phase increases. The dashed line visualises the behaviour $\sigma_\phi \propto \mu^{-1/4}$ which is expected in the absence of longitudinal phase fluctuations. (b) For chemical potentials larger than 8 kHz thermal atoms prevent good contrast. The visibility stays constant for smaller chemical potentials.

We find an opposite behaviour for the fringe contrast which decreases with increasing chemical potential. A maximum value is achieved for $\mu \approx 6\,\text{kHz}$ which is stable under further reduction of the atom number. Decoherence due to longitudinal phase fluctuations seems to be too small to affect the contrast without hold time inside the double well. The contrast drop at higher chemical potential is probably caused by a larger fraction of thermal background atoms. The comparison of the two analysed parameters suggests that the initial BEC should be prepared at chemical potentials between 6 kHz and 8 kHz for best performance of the interferometer.

6.6 Conclusion

In this chapter we demonstrated how two independent BECs are overlapped in free fall and a relative phase measurement can be extracted from the resulting interference pattern. From repeating the same experiment over and over again we built up a statistic in order to prove that the splitting and recombination processes are coherent.

The detailed analysis, however, showed that the release of the atoms is not ideal. The atoms experience a force during release leading to a correlation between the position of the interference pattern and the relative phase. The problem can be overcome by manipulating the projection of the atomic spin while turning off the rf dressing field. This reduces the variation of the observed correlation and a representative phase measurement without correction is possible.

The dephasing of the two double wells sets an upper limit for the interaction time of the interferometer. After a certain hold time the width of the distribution is too large to extract a precise phase measurement. The growth of the phase spread is caused by an uncertainty in the difference of the chemical potential $\sigma_{\Delta\mu}$. Therefore a single measurement of an external signal of interest with the interferometer is afflicted with an uncertainty of 26 Hz, regardless of the interaction time. With p repetitions this limit can be reduced to $26\,\text{Hz}/\sqrt{p}$. For comparison, the gravitational potential of a ^{87}Rb atom increases 2.4 Hz when it is lifted through a height of 1 nm. In one single shot the interferometer with a height separation of 100 nm should measure the weight of an atom with a precision of around 10%.

Another limit is caused by the loss in contrast which plays, however, a minor role compared to the dephasing. As one reason for the loss in contrast we identified the presence of longitudinal phase fluctuations. The crucial point of these fluctuations is not the loss in contrast but rather the enhancement of the phase spreading rate causing a reduced sensitivity of the interferometer.

However, we demonstrated that distinctive phase measurements with quasicondensates are possible. Decoherence sets an upper limit to the time the condensates can spend in the interferometer arms. Our apparatus therefore passed a feasibility study for building an interferometer actually performing a measurement of an applied phase shift.

Chapter 7

Running the Interferometer

In the last chapters we considered the requirements which are essential for an atom chip BEC interferometer. We showed that we split the BEC coherently, that it is possible to determine the relative phase and we explored the limits of such a measurement. Furthermore, we implemented and characterised an asymmetric double well potential. We discussed the feasibility of each step and are now at the point where we can put everything together necessary for the working apparatus: (1) state preparation, (2) coherent splitting, (3) application of a phase shift on one interferometer arm, (4) coherent recombination and (5) detection of the relative phase. What is still missing, however, is the development of a complete scheme combining the four different steps to a working device. The goal is to implement an experimental procedure that allows an independent, interferometric measurement of a tiny external force.

The crucial point in the operation scheme is the application of the phase shift. In general one can apply an external potential gradient of magnetic or electric origin. This would not require any movement of the BECs apart from splitting and recombining and the precision of the measurement would only be limited by the dephasing process.

In this thesis, however, we focus on the measurement of the local gravitational field. In order to achieve a phase shift we introduce a height difference between the interferometer arms. Since the movement relative to the chip surface also generates a shift in the magnetic potential energy, the precision of our device is additionally influenced by the systematic error in the measurement of this effect.

We present two possible interferometer schemes in section 7.1. Finally, in section 7.2 we demonstrate the capabilities of our apparatus by making an absolute measurement of the gravitational acceleration g. The proof that the interferometer works is the main result of this thesis.

7.1 Schemes for a Gravitational Gradient Interferometer

The basis for an interferometric measurement is the preparation of the input state. If the input state is not well known and not repeateable, a prediction for the outcome of the relative phase is impossible. We have already discussed the necessity of a coherent splitting process. It is, however, also important to think about the procedure on how the tilting is implemented in a possible scheme. The tilt induces an energy shift ΔV between the two wells and hence a relative phase shift according to the integral over the duration t of the experimental sequence

$$\phi = -\frac{1}{\hbar} \int_0^t \Delta V(\gamma(\tau), \tau) \, d\tau, \tag{7.1}$$

where the tilt angle $\gamma(\tau)$ is a function in time. The starting point of the integration $t = 0$ is defined by the point in time where the splitting process is complete and no link is left between the two clouds.

In principle there are two possibilities to prepare an energy imbalance in the experiment. One is to split the BEC into a symmetric state and then change the energy difference afterwards. This scheme has the advantage that we always start with the same, well-defined initial relative phase. In the second case the BEC is split directly into an asymmetric double well. But the initial phase state after ramping up the potential barrier is then already dependent on the chosen asymmetry. In our prediction for a measurement outcome we have therefore to include the phase evolution during the complicated process of splitting a BEC into two wells. We will see in the following discussion, however, that this method has the capability of gaining some extra information on the splitting process itself.

7.1.1 Symmetric Preparation

Procedure

In the case of symmetric preparation the linear rf ramp rises a barrier and splits the initial BEC into a symmetric double well potential. The complete scheme for symmetric preparation is presented in figure 7.1. After the splitting process has finished the potential is rotated linearly to a certain tilt angle γ within an interaction time t_{int}. During the movement the energy gap ΔV increases linearly. In order to make sure that there is no phase shift due to the release from the new position, we move the clouds back to their initial position within the same time t_{int}. No additional hold time is applied between the two movements. The energy shift ΔV decreases linearly till the potential is symmetric again and the atoms are released from the trap. Absorption imaging then delivers a fringe pattern after recombining the clouds in free fall. We mention that it is not essential that the double well is symmetric, rather than that the initial well state is always the same after the splitting process. If there is no energy difference, however, we end up with two equal clouds resulting in a high fringe contrast even after tilting, since the link between the BECs is broken and a particle exchange is suppressed.

Relative Phase

We calculate the phase evolution for this scheme by solving the integral of equation (7.1). The potential difference

$$\Delta V = \Delta V_{grav} + \Delta E_{mag} \tag{7.2}$$

consists of the two contributions from gravity and the magnetic potential shift. Since we assume that the clouds are of equal size, atom-atom interactions do not contribute to ΔV. Both are linear in

1. Input State

2. Coherent Splitting

3. Phase Evolution

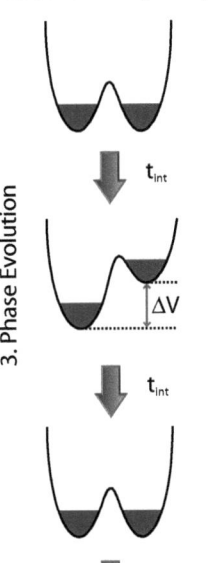

t_{int}

ΔV

t_{int}

4. Coherent Recombination

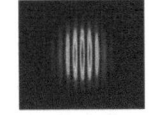

5. Phase Measurement

Figure 7.1: The experimental procedure for an interferometer with an induced height difference using the symmetric preparation scheme. After preparation of the BEC in the magnetic trap we split it into a balanced double well potential. The two wells contain equal atom numbers which cancels the contribution of atom-atom interactions to the relative phase. In the next step we introduce a potential difference ΔV between the interferometer arms by moving the axis of the double well out of the horizontal. The tilting is performed linearly within an interaction time t_{int} which moves one cloud closer to the chip, the other further away. In order to always release from the same position independent of the tilt angle γ we reverse the movement within the same time t_{int}. No additional hold time is applied between the two steps. Finally, we can observe the fringe pattern after turning off the trap and recombining the clouds in free fall. The measured relative phase is the integral over the tilting process.

the angle $\gamma(t)$, hence ΔV shows a linear behaviour as well. Furthermore, ΔV has no explicit time dependence, which means $\partial \Delta V/\partial t = 0$, and $\gamma(t)$ is linear in time. These considerations yield for the relative phase the result

$$\phi(\gamma) = -\frac{1}{\hbar}\Delta V(\gamma) t_{\text{int}}. \tag{7.3}$$

The simple result for the phase shift is linear in both the potential imbalance and the interaction time t_{int}. Independent of these parameters, the clouds are always prepared in the same state and fall under equal conditions from the same position. Hence, phase shifts due to differences in preparation and release position are not expected.

Ideally we find for a balanced double well that $\Delta V(\gamma_0) = 0$ where γ_0 defines the tilt angle at this symmetric configuration. Since it is not clear how good our balancing procedure is and the absolute gravitational potential cannot be determined from the angle measurement, the resulting relative phase has an offset of $\phi_0 = -\frac{2}{\hbar}\Delta V(\gamma_0) t_{\text{int}}$ which is independent of a variation in γ. Observing the behaviour of the phase with tilt angle therefore corresponds to a measurement of the change in ΔV which is

$$\frac{\partial \phi(\gamma)}{\partial \gamma} = -\frac{1}{\hbar}\left(mg\frac{\partial d(\gamma)}{\partial \gamma}\sin\gamma + mgd(\gamma)\cos\gamma + \frac{\partial \Delta E_{\text{mag}}(\gamma)}{\partial \gamma}\right)t_{\text{int}}. \tag{7.4}$$

In the case of small tilt angles we can assume that $\cos\gamma \approx 1$ and $\sin\gamma \approx \gamma$ which causes the first term in the bracket to play a minor role. Since the splitting distance d is not independent of the tilt angle, the gradient $\partial \phi(\gamma)/\partial \gamma$ is not constant. If the variation in d is negligible and since we expect a linear behaviour of the magnetic potential part, the acceleration of free fall can be approximated by

$$g = \frac{p\Lambda}{2\pi M t_{\text{tof}}}\left(\frac{1}{t_{\text{int}}}\frac{\partial \phi(\gamma)}{\partial \gamma} - \frac{1}{\hbar}\frac{\partial \Delta E_{\text{mag}}(\gamma)}{\partial \gamma}\right), \tag{7.5}$$

where Λ is the fringe wavelength in pixel, p the pixel size projected onto the incident imaging beam, t_{tof} the time-of-flight and M the magnification.

Another method is to vary the interaction time while keeping the tilt angle constant. Writing the phase function of equation (7.3) in dependence of the interaction times a balancing offset $\Delta V(\gamma_0)$ changes the function's slope and can be determined from the data. The interferometer is therefore not only able to measure external fields but provides also an additional method to adjust the performance of the beam splitter.

7.1.2 Asymmetric Preparation

The previous method involved a lot of movement, which is acceptable as long as it is adiabatic. In fact it would be much easier to prepare the atoms already in the tilted position and then wait for an interaction time t_{int} before interfering the clouds. Such a procedure would yield the same phase evolution as described by equation (7.3). But it is then necessary to split the BEC directly into the asymmetric double well as sketched in figure 7.2. After a short hold time t_{hold} inside the potential we then observe the interference pattern. Although for a range of small energy gaps imbalanced splitting yields high enough fringe contrast to allow the observation of a fringe pattern, the contribution of atom-atom interactions to the phase due to the different atom number of the clouds has to be considered. Furthermore, the link between the sites of the potential breaks before the splitting process has finished. As soon as the clouds become independent their relative phase starts to evolve while ΔV is still changing. These processes increase the complexity of the scheme influencing both the initial relative phase after preparation as well as the phase gain during the process.

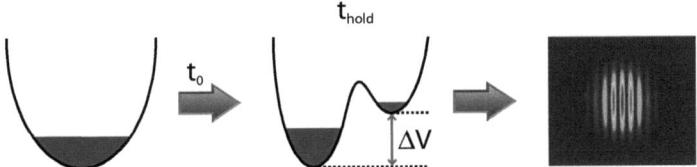

Figure 7.2: Interferometry with asymmetric preparation. The input state is directly split into an asymmetric double well within the time t_0. Therefore the relative phase starts evolving during the splitting process. Further phase evolution is accumulated during the hold stage of the time t_{hold}. The BECs are recombined by release from the asymmetric potential.

The Splitting Process

The potential difference is ramped up linearly during the splitting process to a maximum value $\Delta V(\gamma)$ within a ramping time t_0. $\Delta V(\gamma)$ is well known and linear in γ which we have already proven experimentally. With this knowledge the resulting phase shift is calculated from equation (7.1) to

$$\phi_{\text{split}} = -\frac{1}{\hbar} \int_0^{t_0} \frac{\Delta V(\gamma)}{t_0} t \, dt = -\left[\frac{1}{2} \frac{\Delta V(\gamma)}{t_0 \hbar} t^2 \right]_0^{t_0}. \tag{7.6}$$

This model has to be extended by the fact that two sufficient close BECs with chemical potentials μ_1 and μ_2 adjust each other by tunneling such that $\mu_1 = \mu_2$. Assuming the Thomas-Fermi approximation we find that

$$0 = \mu_a - \mu_b = \Delta V + g(N_a - N_b). \tag{7.7}$$

It follows immediately that a phase shift caused by a potential difference is canceled by the contribution of the mean field energy. Hence in equation (7.6) we have to integrate only over a time $\Delta t = t_0 - t_{\text{split}}$, where t_{split} is the point in time at which the two clouds are separated by a distance d_{split} and the tunnel coupling breaks down. We can then write

$$\phi_{\text{split}} = -\frac{1}{\hbar} \int_0^{\Delta t} \frac{\Delta V(\gamma)}{t_0} t \, dt = -\frac{1}{2} \frac{\Delta V(\gamma)}{t_0 \hbar} \Delta t^2 \tag{7.8}$$

$$\Rightarrow \phi_{\text{split}} = -\frac{1}{2} \frac{\Delta V(\gamma)}{t_0 \hbar} \left(t_0 - \frac{d_{\text{split}}}{d} t_0 \right)^2, \tag{7.9}$$

with d the final splitting distance. We know that according to the implementation of the asymmetric potential the wavelength Λ depends on γ which we approximate linearly by $\Lambda(\gamma) = k\gamma + \Lambda_0$. Using the well known relation between splitting distance and wavelength according to equation (2.21) we write

$$\phi_{\text{split}} = -\frac{1}{2}\frac{\Delta V(\gamma)}{\hbar}t_0\left(1 - d_{\text{split}}\frac{m}{\hbar t_{\text{tof}}}\Lambda_0 - d_{\text{split}}\frac{m}{\hbar t_{\text{tof}}}k\gamma\right)^2, \qquad (7.10)$$

which contains terms of $\propto \gamma^3$.

The Hold Time

After the ramp of the potential barrier has finished the atoms are held inside the two interferometer arms. The phase evolution is now proportional to the hold time t_{hold} because $\Delta V(\gamma)$ is not further varied. However, an essential part of $\Delta V(\gamma)$ is still suppressed by the difference in mean field energy, because the BECs are of unequal size. From equation (7.1) and the discussion of the previous section follows

$$\phi_{\text{hold}} = -\frac{\Delta V(\gamma)\Delta t}{\hbar t_0}t_{\text{hold}} \qquad (7.11)$$

$$\Rightarrow \phi_{\text{hold}} = -\frac{\Delta V(\gamma)}{\hbar}\left(1 - d_{\text{split}}\frac{m}{\hbar t_{\text{tof}}}\Lambda_0 - d_{\text{split}}\frac{m}{\hbar t_{\text{tof}}}k\gamma\right)t_{\text{hold}}. \qquad (7.12)$$

We find terms of the phase shift $\propto \gamma^2$ with a quadratic behaviour in tilt angle. In order to determine ΔV, however, either the point in time or the splitting distance at which the wells lose their tunnel coupling have to be well-known. Both are not directly accessible and therefore the scheme is not ideal for an interferometric measurement of external fields. On the other hand, if the energy gap of the asymmetric potential is characterized, this type of experiment allows us to make a conclusion on the splitting process itself.

One should also notice that our discussion has also impact on our conclusions of the symmetric preparation scheme. The measurement of the potential offset in an experiment where the interaction time is varied leads to an underestimated value of $\Delta V(\gamma_0)$. The imbalance generates a difference in mean field energy between the wells during splitting. According to equation (7.7) the mean field contribution has the opposite sign with respect to the potential offset and hence cancels a part of the phase evolution caused by $\Delta V(\gamma_0)$. Although the interferometer with symmetric preparation is sensitive to variations in asymmetry, the determination of an absolute energy gap requires a complete understanding of the balancing and splitting process itself.

7.2 The Gravitational Gradient Interferometer

The definition of an operation scheme for the interferometer not only provides a model to interpret the outcome of a measurement it also enables us to make a comparison with the characterisation of the double well potential which provides a transformation from tilt angle to a potential difference between the wells. The direct comparison will lead us to the measurement of an external field namely the acceleration of free fall g.

Finally, we arrived at the point where we can make a prediction on the phase evolution of the interferometer based on the equations (7.3), (5.31) and the measured magnetic energy shift. We now arrive at the final step and run the interferometer.

7.2.1 Measurement of the Balancing

In a first experiment we measure the relative phase against interaction time t_{int} using the symmetric preparation scheme as described in section 7.1.1. We prepare the double well initially at an asymmetry of $\alpha = 0$ V which corresponds to an angle $\gamma_0 = -1.05°$ followed by the tilt up and down. The phase evolution is linear in the interaction time and its slope depends on the potential difference $\Delta V(\gamma)$ according to equation (7.3). We take data sets at three different asymmetries, namely $\alpha = 3$ V, -3 V and -5 V where each data point represents 20 repetitions. The first two imbalances are of great interest, as the results for tilts of equal amount in angle but opposite direction are expected to differ only in sign. We have to mention that the data for $\alpha = -3$ V were taken with the same parameters but on a different day which could cause a small change in the trap bottom. The fits presented in figure 7.3 yield slopes of 3.05 ± 0.36 rad/ms, -3.63 ± 0.31 rad/ms and -5.16 ± 0.71 rad/ms, respectively. The change from positive to negative slope indicates a change of sign in the potential difference $\Delta V(\gamma)$ as expected.

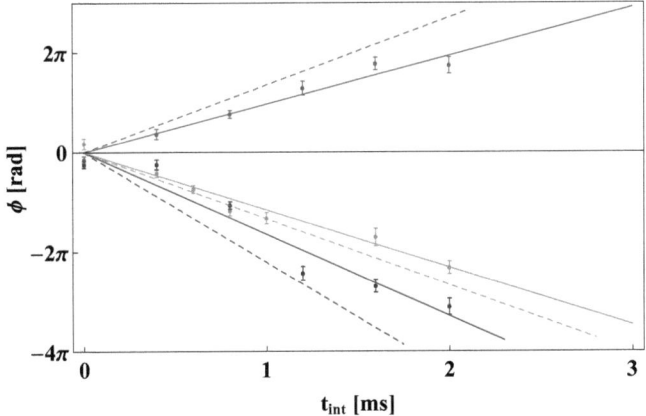

Figure 7.3: Phase evolution with interaction time t_{int}. As the potential changes sign the phase shift changes its sign. Grey: $\alpha = 3$ V, light grey: $\alpha = -3$ V and dark grey: $\alpha = -5$ V.

The whole purpose of the balancing scheme was to achieve a 50 : 50 beam splitter which is important in order to prepare two equal input states for the interferometer. Instead of counting the atoms in the wells or looking for the highest contrast of the fringe pattern the interferometer offers a method to compare the slopes at different asymmetries. Considering the results at $\alpha = 3$ V and -3 V suggests that the double well potential has a small, initial offset in potential difference $\Delta V(\gamma_0) \neq 0$. A calculation yields a balanced double well for an asymmetry parameter of $\alpha = -0.27$ V. Taking all three determined slopes into account we conclude, however, that the potential is well balanced within the error of the fits.

Also presented in figure 7.3 is the expected phase evolution assuming a zero energy gap at an angle γ_0 and the knowledge of the gravitational acceleration g. An initial potential offset would change the slope according to equation (7.3) and corresponds basically to a rotation of the data around the origin at $t_{int} = 0$. The slopes in the model have values of 4.26 rad/ms, −4.19 rad/ms and −6.95 rad/ms at the considered asymmetries respectively. We find that the measured phase evolution is smaller than expected and we have to remember the discussion on the asymmetric preparation scheme. The small, suggested initial imbalance could cause a reduced phase evolution due to mean field interaction. Since the data sets were not taken on the same day, a change in this parameter could further complicate the situation. The deviation of the slopes from the expectation lies between 14% and 29%. The found imbalance comparing the data at $\alpha = -3$ V and 3 V would suggest a correction of the slope of up to 10%. Therefore other effects must contribute to the measurement error. Especially for interaction times longer than 1.5 ms the data points seem to disagree with the model which could be due to dephasing effects.

7.2.2 Measurement of the Acceleration of Free Fall

In the next step we use again the symmetric preparation scheme as presented in figure 7.1. But instead of varying the interaction time we measure the phase as a function of the tilt angle. This method has two advantages compared to the last experiment. First, the gain in relative phase between two data points is independent of a possible energy gap offset at $\alpha = 0$ V. Second, because of the fixed interaction time the phase spread is constant for all taken data points.

According to the discussion in section 6.5.1 we choose the total time that the atoms spend in the interferometer arms to a maximum of 4 ms. Hence we repeat the experiment at three fixed interaction times, 1 ms, 1.6 ms and 2 ms. The results are presented in figure 7.4 which shows a linear change in phase with tilt angle γ. Hence we conclude that the energy shift between the two wells behaves linearly according to our expectations. The plotted lines in figure 7.4 represent the expected relative phase calculated from the gravitational shift and the determined magnetic energy shift for each interaction time respectively. The data points agree well with the expectation of the model. Linear fits to the data points yield slopes of 1.45 ± 0.04 rad/° for $t_{int} = 1$ ms which is nearly half the value of 2.69 ± 0.05 rad/° at an interaction time of 2 ms.

The slope of the phase evolution is a direct measure of the potential gradient between the two wells and allows the measurement of an external field. In the following we assume that we have no knowledge about the external gravitational acceleration field except that it is orientated along the z axis. The linearity of the data implies immediately that the force is homogenous over the extension of the double well. In order to determine the value of g we fit a model according to equation (7.3) to the three data sets. The procedure delivers values for g which lie within a range of 10% to the expected value of 9.81 m/s². The measured values for the three interaction times are 10.59 ± 0.80 m/s², 9.98 ± 0.52 m/s² and 8.66 ± 0.57 m/s², respectively. The error can be reduced by calculating the mean value yielding the final measurement result for g of $9.7\,\text{m/s}^2$ with a statistical error of $\pm 0.5\,\text{m/s}^2$ or 5%.

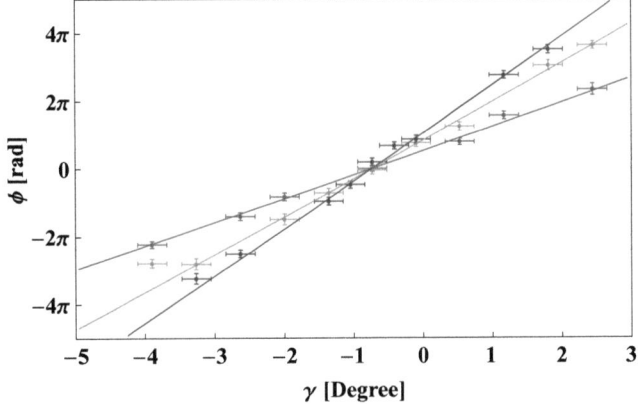

Figure 7.4: The figure presents the relative phase versus tilt angle γ. The data sets correspond to three different interaction times t_{int}. Grey: $t_{\text{int}} = 1\,\text{ms}$, light grey: $t_{\text{int}} = 1.6\,\text{ms}$ and dark grey: $t_{\text{int}} = 2\,\text{ms}$. The plotted lines represent the expected relative phase calculated from the measurements on the gravitational and magnetic energy shift.

7.3 The Asymmetric Interferometer

Although we have successfully implemented an interferometer by splitting the BEC into a symmetric double well followed by a rotation of the potential, we continue with the asymmetric prepartion scheme whose phase evolution is more complicated. The last experiment gave us information about the potential gradient between the two wells which we verified by measuring the gravitational acceleration. The gained knowledge reduces the unknown parameters in the equations (7.10) and (7.12) on the asymmetric preparation scheme to the splitting distance d_{split} at which the two wells lose their coupling. By implementing this extended scheme we can therefore obtain additional information about our system.

7.3.1 Influence of the Splitting Process

Tunneling is a well-known phenomenon predicted by quantum mechanics. First studies predicted the Josephson effect [122] between two superconductors separated by a thin insulator. In recent years the observation of Josephson oscillations was suggested for two weakly coupled BECs [123]. The first experiment [124] on such a system indeed verfied an oscillation in population imbalance and phase which is not able to evolve independently.

The splitting of the BEC in our experiment is a dynamic process where the system is transformed from one single BEC via the Josephson regime into two independent BECs. The transition is smooth and not well defined but as soon as the link between the BECs becomes weak enough the phase will start to evolve independently. The raising of a potential barrier and the behaviour of the system was described theoretically by different models [70, 71, 125]. In this context a popular method is to divide the ramping of the barrier into two parts, where the BECs are still coupled and another one where they are uncoupled, and treat each part independently. An experimental *in situ* observation of the coupling loss, however, is difficult, since the process is dynamic. Our method promises to deliver a number for the coupling distance d_{split}. But we remark that our experimental derivation of g was not a high precision measurement and so will not be the more complicated scheme which will additionally be influenced by statistical errors of the last experiment. Moreover, at this point we set value on the order of magnitude of the measurement and the proof of our theoretical considerations on the preparation scheme.

7.3.2 Measurement of the Coupling Length

In the following we implement the double well with exactly the same parameters as in the previous measurements but the procedure for the interferometer is different. The BEC is directly split into an asymmetric double well by ramping the modulation voltages from 0 V to the according values. Afterwards we hold the clouds for a certain time t_{hold} and release from the tilted position. We expect that the release from rotated positions changes the wavelength of the fringe pattern, however, the phase measurement will not be affected. Again we calculate the mean phase at each data point from 20 repetitions.

The plot in figure 7.5 presents the results for the hold times $t_{\text{hold}} = 1$ ms and 2 ms. The phase evolves clearly nonlinearly. Neglecting the asymmetric splitting process and considering only the phase gain from the time the atoms spend inside the interferometer arms, we would expect a behaviour described by equation (7.3) with t_{int} replaced by t_{hold}. A linear fit, however, shows slopes of around 0.87 rad/° and 1.30 rad/°, respectively, and is strongly reduced in comparison to the result of the symmetric scheme.

According to the theory the total phase shift $\phi = \phi_{\text{split}} + \phi_{\text{hold}}$ is composed of the part acquired during the splitting process and the one gained during the hold time in the interferometer arms. We fit a model of ϕ to the data sets assuming a linear evolution of the energy shift ΔV according to the interferometer measurements of the last section. As a fit parameter we use the splitting distance d_{split} where the phase starts to evolve independently. The fits are shown as solid lines in the plot. We find a value of $d_{\text{split}} = 2.66 \pm 0.20\,\mu\text{m}$ at $t_{\text{hold}} = 1$ ms while the fit for 2 ms hold time yields $d_{\text{split}} = 2.32 \pm 0.50\,\mu\text{m}$.

From the fringe wavelength within the operational range of the interferometer between tilt angles of around $-4°$ and $2°$ we found that the clouds are independent at least down to a splitting distance of $3.05\,\mu\text{m}$. We expect the phase to be locked at well separations where only one single fringe is visible in the interference pattern. An exact comparison is not possible at this point as the simple formula does not hold for large wavelengths and a conversion between rf current and well separation is not available. However, a rough estimation from the simple relation between wavelength and well separation yields a splitting distance of $1.9\,\mu\text{m}$. The measurement suggests that the link breaks at slightly larger separations which is reasonable because also after the wavefunctions lost their overlap tunneling through the barrier can still be possible. Although the chemical potential of each BEC is smaller than the barrier, the phase can still be locked at small enough barrier widths.

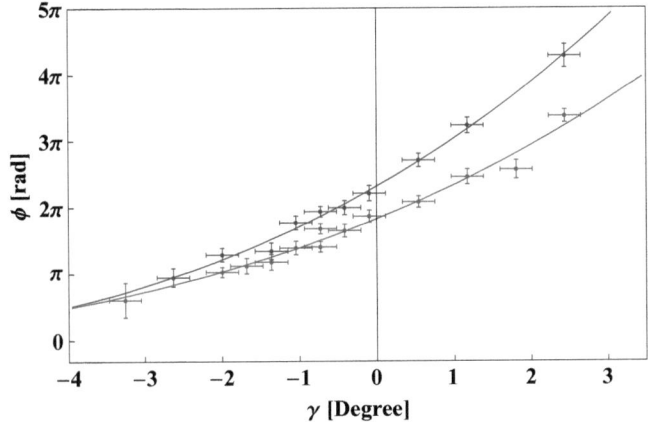

Figure 7.5: Relative phase measured against tilt angle using the asymmetric scheme. The range over which the phase varies is clearly reduced compared to the symmetric procedure. The phase evolution is non-linear and the solid lines are a fit of our theoretical model. Grey: $t_{hold} = 1$ ms, dark grey: $t_{hold} = 2$ ms.

7.4 Conclusion and Error Discussion

We discussed two possible schemes for running the interferometer and derived the corresponding equations for the evolution of the relative phase. Concerning the measurement of the acceleration of free fall the symmetric preparation scheme yields simple equations and has the advantage that the atoms are always released from the same position. In the case that we monitor the relative phase with fixed interaction time and the tilt angle as a variable, the measurement is even independent of an initial imbalance.

Error of the g Measurement

The experiment delivered a value for g close to the expected value of 9.81 m/s^2 with a statistical error of around 5%. We identify the origin of this statistical error with the uncertainty in the phase from shot-to-shot.

The question about the absolute precision with which we can determine the acceleration of free fall, however, is also based on the systematic error introduced in the characterisation of the double well potential which was presented in chapter (5). Hence, to cover the question on how precise this result really is we have to make a rigorous error treatment.

We derive g from equation (7.4). The errors propagating the measurement consist of the uncertainties on the magnification, the wavelength, the tilt angle, the slope of the magnetic potential shift and the phase measurement. An overview of the errors on these parameters which we found in earlier sections is given in table (7.6). Taking all uncertainties into account we expect to find a value for g

Parameter	Uncertainty [%]	Relative Error in g [%]
Magnification M	2	2
Fringe Wavelength Λ	1	1
Tilt Angle $\partial \gamma / \partial \alpha$	6	6
Magnetic Potential Shift $\partial \Delta E_{mag} / \partial \gamma$	9	13
Phase Shift $\partial \phi / \partial \gamma$	4	8
Total Uncertainty on g		16

Figure 7.6: Uncertainties on the different parameters of the g measurement. The uncertainty on the acceleration of free fall is derived from Gaussian error calculation. We identify that the largest fraction on this error is caused by the uncertainty on the phase measurement and on the magnetic potential difference.

to within 16%. The main contributions come from the dephasing and the large systematic error in the magnetic energy shift. The latter is especially important, since we expect the magnetic energy shift to be around double the amount of the gravitational potential difference. The error due to magnetic coupling sets therefore a limit to the precision of this type of interferometer where the orientation of the double well is tilted with respect to the chip surface. Although a phase measurement with the interferometer might be precise, the analysis of an external field is limited by the characterisation of the double well potential. As already mentioned, the single values of g found for the three different interaction times lie all within a 10% range around the expectation. This observation is consistent with our error estimation.

Regarding the three major contributions to the error, a larger amount of taken data points would significantly decrease the uncertainty. Especially a more precise measurement of the shift in magnetic energy could drop the total error below 10%.

Our error estimation, however, does not take long term fluctuations into account. The data needed in order to arrive at a g measurement was taken over several weeks. We mentioned already, for example, that the trap bottom showed some fluctuations which would also influence the splitting distance. We estimated that the contribution of the known day to day fluctuations are neglegible compared to the uncertainties on the various parameters. Such variations, however, will be of interest as soon as the interferometer will achieve precision measurements on the basis of only a few percent.

Conclusion on Splitting Dynamics

Using the asymmetric scheme we could gain additional information about the splitting process and its phase dynamics. The experiment showed qualitative agreement with theoretical considerations. During the rise of the barrier in the middle of the trap, the two wells lose their coupling and the relative phase starts to evolve while the potential difference still increases and the phase has no longer a linear behaviour. This behaviour was also studied by Schumm *et al.* [13] who monitored the phase while varying the correlated parameters splitting distance and time. We, however, study the relative phase with a change in asymmetry and splitting distance but constant splitting time. The crucial point is that atom-atom interactions counteract the asymmetry in the potential and are responsible for a reduced phase evolution. Although we could extract numbers for the splitting distance where the coupling is lost, we expect an uncertainty of more than 16%. We believe, however, that by improving the interferometer precision we could make exact measurements to examine the splitting process and deliver a useful, experimental comparison for the intensive, theoretical studies.

Chapter 8

Conclusions and Outlook

8.1 Summary

In this thesis we described in detail how an atom BEC interferometer can be implemented on an atom chip. The starting point is an atom chip with four simple Z-wires. Together with an external bias field the atom chip creates an IP trap near its surface in which a BEC is prepared.

In order to implement the interferometer we need to split the BECs representing the input state into two independent arms. Therefore we make use of a very popular and widespread technique which uses rf fields to dress Zeeman states of the atoms. The dressed atoms then experience a new potential environement having the shape of a double well. We successfully implement rf adiabatic potentials in our system by overlapping the two dc chip wire currents with two independent rf signals.

After separating the atoms in two wells we recombine them in free fall and we determine their relative phase from the resulting interference pattern. We conduct a detailed analysis on the coherence of the wells and identify a main limitation due to phase spreading. The precision of an interferometric measurement depends on the interaction time the atoms spend in the interferometer arms which is set to a maximum of 5 ms in our setup.

The feasibility of a phase measurment and hence the measurement of an external field by just splitting a BEC and recombining the two parts was demonstrated by several groups in the past. The point which makes our interferometer distinctive, however, is its operation scheme and the way we introduce a potential difference between its two arms.

The independent control of the rf currents allows us to dynamically change the orientation of the rf field vector while keeping the rf field strength nearly constant. Theoretical predictions expect the tilt of the double well with respect to the horizontal at the same time which we confirm experimentally. The consequence of the tilt is that one BEC cloud is closer to the chip surface than the other introducing a relative potential difference due to gravity and rf coupling strength. Using this technique we create a double well potential whose asymmetry in energy can be tuned over a certain range. In order to make predictions for the interferometer we conduct a detailed analysis of the energy gap. We determine the gravitational contribution by mapping the splitting distance and the tilt angle for a certain set of rf current configurations. By a spectroscopic analysis we obtain the difference in magnetic coupling between the two wells. Here we find that the shift due to magnetic coupling is around twice the shift due to the gravitational difference.

Finally, we implement an interferometer by putting the preparation of the BEC, splitting the BEC, tilting the double well potential and recombining the two wells into one experimental scheme. Theoretical considerations predict that the acceleration of free fall can be calculated from the slope of the relative phase with the tilt angle $\partial \phi / \partial \gamma$. An error analysis of the various parameters yields an unceraintly on such a measurement of around 16%. Indeed we derive values for g from the experiment within this interval and an statistical error of 5%. Especially, for this type of interferometer the main contribution to the uncertainty is given by the systematic error on the outcome of the rf spectroscopy and the phase spreading. First propagates the calculation of g from the measured phase while second is mainly responsible for the statistical error.

We successfully built an atom chip BEC interferometer using a completely new scheme to introduce a relative phase shift. We demonstrated that our apparatus is capable of making an absolute measurement of the acceleration of free fall and identified the limits of the system.

8.2 Outlook

By proving the principle of operation and the feasibility of measuring independent, external fields the goal is now to improve the precision and capability of the interferometer. This can only be achieved with strategies for reducing the three main uncertainties in the experiment.

One systematic error we make is the measurement of the angle during free fall. A more precise result would be achieved by observing the orientation of the double well *in situ*. Therefore, we currently increased the magnification of our imaging system by a factor of three. Since this might still not be enough to resolve the double well, we also want to switch to rf oscillators with the capability of a phase locked frequency sweep which is currently not possible. Splitting a BEC by sweeping the rf frequency instead of using high rf field strengths allows the implementation of much larger well separations.

We emphasised already the role of the error on the rf spectroscopy results. This does not pose a problem in the measurement of external fields, i.e. magnetic or electric field gradients, which do not require tilting of the double well potential. In the case of the gravity measurement a symmetrical geometry, for example mounting the chip vertically and splitting parallel to the surface, has the same effect. In the current setup, however, a much more careful measurement of the magnetic coupling and the collection of more data points is necessary. Especially, the amount of experiments for deriving some numbers is immense for spectroscopy. The data set presented in this thesis already consists of around 300 single measurements which should be taken within one single day. Therefore, also the reduction of the experimental cycle time is an issue.

The most promising strategy to reduce the phase spreading is the implementation of number squeezing. We hope to achieve new atomic double well states by adjusting the shape and the speed of the rf ramp. We did not analyse the influence of the rf ramp on the phase states throughout this thesis but we believe that a detailed examination will be worth the effort. In the past a dramatic reduction of the dephasing rate by squeezing the number difference was observed in sodium [15]. It was shown that the uncertainty in relative chemical potential $\sigma_{\Delta\mu}$ was reduced by a factor of 10. Alternatively, we could implement a less extreme trap geometry or lower temperatures of the BEC to avoid longitudinal phase fluctuations. This would slow down dephasing by a factor of $\sqrt{2}$ in our case. Furthermore, it is possible to reduce interactions by a Feshbach resonance, as demonstrated, for example in [126] using ^{39}K. With some combination of these measures one can expect a fundamental noise level of $\sigma_{\Delta\mu} \approx 1$ Hz.

With exception of mounting the chip vertically all other improvements on the systematic errors can be implemented by changing the control on external magnetic fields. Such extensions should be easily made to an existing experiment and should already improve the absolute measurement precision by several percent. By cancelling the spatial variation of the rf field strength, the most troublesome systematic error, we expect the absolute precision of our device to improve down to 1%.

The small size of the atom cloud and its proximity to a surface make trapped BEC interferometry attractive for mapping atom-surface interactions. For example, a Rb atom 1 μm from a plane conductor has a Casimir-Polder interaction energy of 270 Hz [127], decreasing to 3.3 Hz at a distance of 3 μm. Over this range, it should be possible to make measurements with a few per cent accuracy, limited at large distance by the noise level and at short distance by uncertainty in the spatial distribution of the atoms. This offers the possibility of improving over the existing measurements of the Casimir-Polder interaction [128, 129] and of its temperature dependence [130]. The related phenomenon of Casimir attraction between two macroscopic bodies has been measured with 15% accuracy for parallel plates [131] and with precision in the range 1%-5% between plane and curved surfaces [132]. Therefore our type of interferometer is suitable for accurate measurements of atom-suface interactions.

Appendix A

Rubidium

Rubidium is an alkali metal with only one valence electron in the 5S shell. We consider exclusively the isotope ^{87}Rb in this thesis with an atomic mass $m = 1.443 \times 10^{-25}$ kg. ^{87}Rb is slightly radioacitve and decays with emission of β radiation into ^{87}Sr. The lifetime of this decay, however, is 49 billion years making it effectively stable and suitable for experiments with a cycle time in the order of a minute.

The transition of the D$_2$ line from the 5S$_{1/2}$ ground state to the 5P$_{3/2}$ has a wavelength of 780 nm. The availability of diode lasers at this wavelength makes rubidium convenient for laser cooling. The different hyperfine transitions of the D$_2$ line are presented in figure A.1. For cooling we use the $|F = 2\rangle \longrightarrow |F' = 3\rangle$ transition. After the spontanous emission of a photon the atom either ends up in the $|F = 1\rangle$ or the $|F = 2\rangle$ state. In order to achieve a closed cooling cycle a repump laser beam is at the $|F = 1\rangle \longrightarrow |F' = 2\rangle$ transition is overlapped with the cooling beam.

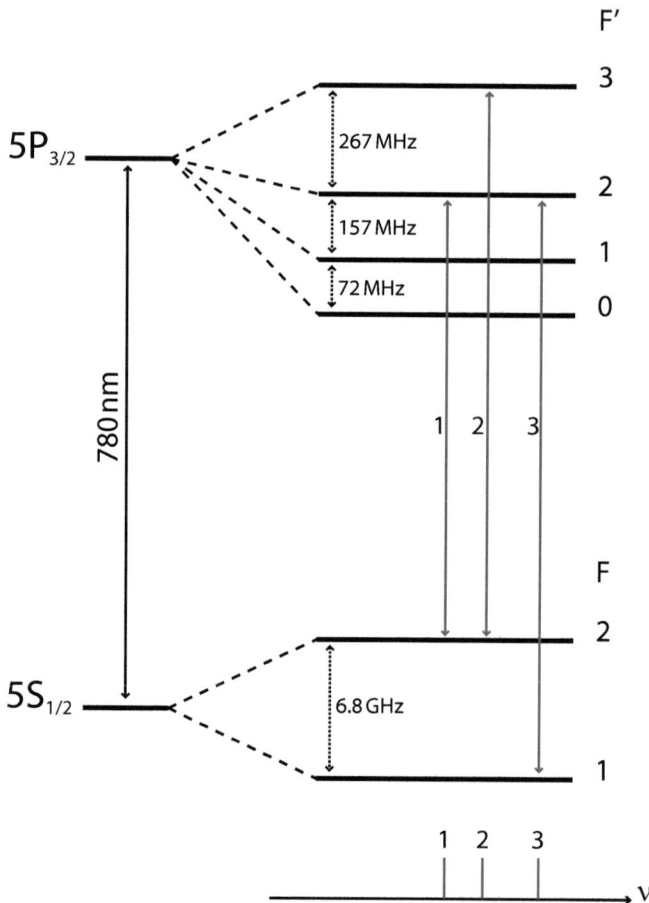

Figure A.1: Schematic diagram of the D_2 line of ^{87}Rb. The $5S_{1/2}$ ground state and the $5P_{3/2}$ excited state split up into the hyperfine states with the F-numbers F and F' respectively. The transitions between the hyperfine levels are used for optical pumping (1), the cooling beam (2), absorption imaging (2) and the repump laser beam (3).

Appendix B

Directional Statistics

B.1 The Mean Direction

A certain set of angles θ_k with $k = 1,2,...,n$ can always be interpreted as a set of directions. In the complex plane we refer to a direction as a unit vector $\mathbf{x}_k$ on a unit circle. In this picture the angles θ and $\theta + 2\pi$ reproduce the same point on the circle. A small distribution of angles around an angle $\theta_0 = 0$ (radians) might clearly have a distinctive direction along θ_0. Defining a sample mean analogue for data on a line would yield a mean angle of $(\theta_1 + ... + \theta_n)/n$. Considering, however, angles of $\pi/4$ and $3\pi/4$, we calculate a sample mean of $\pi/2$ which points in the opposite direction of θ_0. Intuitively we say that the usual sample mean cannot work for data on a circle and a modified concept is needed. Instead we have to define the mean direction $\bar{\theta}$ as a summation of all unit vectors $\mathbf{x}_k$

$$\bar{r}e^{i\bar{\theta}} = \frac{1}{n}\sum_{k=1}^{n} e^{i\theta_k}. \tag{B.1}$$

Since the Cartesian coordinates of the vectors $\mathbf{x}_k$ are $(\cos\theta_k, \sin\theta_k)$, the coordinates $(\bar{C}, \bar{S})$ of the vector in equation (B.1) are calculated according to

$$\bar{C} = \frac{1}{n}\sum_{k=1}^{n} \cos\theta_k \quad \text{and} \quad \bar{S} = \frac{1}{n}\sum_{k=1}^{n} \sin\theta_k. \tag{B.2}$$

The mean resultant length $\bar{r}$ length is then nothing else than

$$\bar{r} = \left(\bar{C}^2 + \bar{S}^2\right)^{1/2} \tag{B.3}$$

and in the case that $\bar{r} > 0$ we find the mean angle

$$\bar{\theta} = \begin{cases} \arctan\left(\frac{\bar{S}}{\bar{C}}\right), & \text{if } \bar{C} \geq 0, \\ \arctan\left(\frac{\bar{S}}{\bar{C}}\right) + \pi, & \text{if } \bar{C} < 0. \end{cases} \tag{B.4}$$

For $\bar{r} = 0$ an angle $\bar{\theta}$ is not defined. This happens, for example, in a set of two unit vectors with angles 0 and $\pi/2$. The two directions are completely opposite and hence a certain direction of the data is not distinctive. A vanishing mean resultant can therefore be interpreted as completely random distribution of vectors on the unit circle.

B.2 Circular Variance and Standard Deviation

For data on the line we have measures like the variance and the standard deviation to compare the spread of a certain data set. For the circular data we found that the mean resultant length describes the concentration of the data. Since the x_i are unit vectors, it is $0 \le \bar{r} \le 1$ where a mean resultant length close to 1 means that the direction of the data is very concentrated. In the literature the circular variance is therefore defined as

$$v = 1 - \bar{r} \tag{B.5}$$

with values in a range of $0 \le v \le 1$. It is also useful to define a standard deviation for circular data. The circular standard deviation is written in terms of the variance as

$$\sigma = \sqrt{-2 \log (1 - v)} \tag{B.6}$$

which takes, analogous to the usual standard deviation, values in the interval $[0, \infty]$. The closer the value of σ is to 0, the smaller is the spread of the data.

B.3 Distribution Functions

Considering distribution functions on a circle we have especially to regard its boundary conditions. Let $F(x)$ be a distribution function which gives the probability that we find an angle θ in the interval between 0 and x then the function F has to match the condition

$$F(x + 2\pi) - F(x) = 1, \quad -\infty < x < \infty. \tag{B.7}$$

Hence the probability to measure a data point within an arc of length 2π on the unit circle is 1. In the case that the function $F(x)$ is absolutely continous, there is a probability density function $f(x) = F'(x)$ with the properties

(i) $f(x) \ge 0$ almost everywhere on $(-\infty, \infty)$,

(ii) $f(x + 2\pi) = f(x)$ almost everywhere on $(-\infty, \infty)$,

(iii) $\int_0^{2\pi} f(x)\, dx = 1$.

The simplest distribution on a circle is the uniform distribution with a probability density $f(x) = \frac{1}{2\pi}$. But it is inadequate to describe data concentrated around a certain direction. Therefore one could wrap a Gaussian distribution around the circle. A key role for circular data, however, plays the von Mises distribution $M(\bar{\theta}, \kappa)$ [120] with a probability density function

$$g(x, \bar{\theta}, \kappa) = \frac{1}{2\pi I_0(\kappa)} e^{\kappa \cos(x-\bar{\theta})}. \tag{B.8}$$

The distribution is symmetrical around the mean direction $\bar{\theta}$. The modified Bessel function of order 0 is defined by the integral

$$I_0(\kappa) = \frac{1}{2\pi} \int_0^{2\pi} e^{\kappa \cos \theta} d\theta. \tag{B.9}$$

The spread of the data is described by the concentration parameter κ which is connected to the mean resultant length via the relation

$$\bar{r} = \frac{I_1(\kappa)}{I_0(\kappa)} \tag{B.10}$$

with $I_1(\kappa)$ the modified Bessel function of order 1. By fitting the von Mises distribution to a data set we can easily determine the parameters $\bar{\theta}$, κ and hence $\bar{r}$. For $\kappa = 0$ equation (B.8) reproduces the uniform distribution and the measured data is completely random.

Bibliography

[1] R. Rojas, "How to make Zuse's Z3 a universal computer," Annals of the History of Computing **20**, 51–54 (1998).

[2] A. A. Michelson, H. A. Lorentz, D. C. Miller, R. J. Kennedy, E. R. Hedrick, and P. S. Epstein, "Conference on the Michelson-Morley experiment held at Mount Wilson," Astrophysical Journal **68**, 341 (1927).

[3] C. Davisson and L. H. Germer, "Diffraction of electrons by a crystal of nickel," Phys. Rev. **30**, 705–740 (Dec 1927).

[4] I. Estermann and A. Stern, "Beugung von molekular Strahlen (bending of molecular rays)," Z. Phys. **61**, 95 (1930).

[5] J. A. Leavitt and F. A. Bills, "Single-slit diffraction pattern of a thermal atomic potassium beam," Am. J. Phys. **37**, 905 (1969).

[6] D. W. Keith, M. L. Schattenburg, H. I. Smith, and D. E. Pritchard, "Diffraction of atoms by a transmission grating," Phys. Rev. Lett. **61**, 1580–1583 (Oct 1988).

[7] C. S. Adams, M. Sigel, and J. Mlynek, "Atom Optics," Physics Reports **240**, 143–210 (1994).

[8] K. P. Zetie, S. F. Adams, and R. M. Tocknell, "How does a Mach-Zehnder interferometer work?." Physics Education **35**, 46 (2000).

[9] L. Marton, "Electron interferometer," Phys. Rev. **85**, 1057–1058 (Mar 1952).

[10] H. Rauch and S. A. Werner, *Neutron Interferometry* (Oxford University Press, 2000).

[11] D. W. Keith, C. R. Ekstrom, Q. A. Turchette, and D. E. Pritchard, "An interferometer for atoms," Phys. Rev. Lett. **66**, 2693–2696 (May 1991).

[12] A. Peters, K. Y. Chung, and S. Chu, "Measurement of gravitational acceleration by dropping atoms," Nature **400**, 849–852 (1999).

[13] T. Schumm, S. Hoferberth, L. M. Andersson, S. Wildermuth, S. Groth, I. Bar-Joseph, J. Schmiedmayer, and P. Krüger, "Matter-wave interferometry in a double well on an atom chip," Nature Physics **1**, 57 (2005).

[14] Y. Shin, M. Saba, T. A. Pasquini, W. Ketterle, D. E. Pritchard, and A. E. Leanhardt, "Atom interferometry with Bose-Einstein condensates in a double-well potential," Phys. Rev. Lett. **92**, 5 (2004).

[15] G.-B. Jo, S. Will Y. Shin, T. A. Pasquini, M. Saba, W. Ketterle, and D. E. Pritchard, "Long phase coherence time and number squeezing of two Bose-Einstein condensates on an atom chip," Phys. Rev. Lett. **98**, 030407 (2007).

[16] R. Grimm, M. Weidemüller, and Y. B. Ovchinnikov, "Optical dipole traps for neutral atoms," Advances in Atomic, Molecular and Optical Physics **42**, 95–170 (2000).

[17] S. Chu, "Nobel Lecture: The manipulation of neutral particles," Rev. Mod. Phys. **70**, 685–706 (Jul 1998).

[18] J. D. Miller, R. A. Cline, and D. J. Heinzen, "Far-off-resonance optical trapping of atoms," Phys. Rev. A **47**, R4567–R4570 (Jun 1993).

[19] D. M. Stamper-Kurn, M. R. Andrews, A. P. Chikkatur, S. Inouye, H.-J. Miesner, J. Stenger, and W. Ketterle, "Optical confinement of a Bose-Einstein condensate," Phys. Rev. Lett. **80**, 2027–2030 (Mar 1998).

[20] J. Fortágh and C. Zimmermann, "Magnetic microtraps for ultracold atoms," Rev. Mod. Phys. **79**, 235–289 (2007).

[21] O. Zobay and B. M. Garraway, "Two-dimensional atom trapping in field-induced adiabatic potentials," Phys. Rev. Lett. **86**, 1195–1198 (Feb 2001).

[22] O. Zobay and B. M. Garraway, "Atom trapping and two-dimensional Bose-Einstein condensates in field-induced adiabatic potentials," Phys. Rev. A **69**, 023605 (Feb 2004).

[23] G.-B. Jo, J.-H. Choi, C. A. Christensen, Y.-R. Lee, T. A. Pasquini, W. Ketterle, and D. E. Pritchard, "Matter-wave interferometry with phase fluctuating Bose-Einstein condensates," Phys. Rev. Lett. **99**, 240406 (2007), http://link.aps.org/abstract/PRL/v99/e240406

[24] S. Hofferberth, I. Lesanovsky, B. Fischer, J. Verdu, and J. Schmiedmayer, "Radiofrequency-dressed state potentials for neutral atoms," Nature Physics **2**, 710–716 (2006).

[25] W. D. Phillips, "Nobel Lecture: Laser cooling and trapping of neutral atoms," Rev. Mod. Phys. **70**, 721–741 (Jul 1998).

[26] W. Petrich, "Ultrakalte Atome: Die Jagd zum absoluten Nullpunkt," Physik in unserer Zeit **27**, 206 (1996).

[27] H. F. Hess, G. P. Kochanski, J. M. Doyle, N. Masuhara, D. Kleppner, and T. J. Greytak, "Magnetic trapping of spin-polarized atomic hydrogen," Phys. Rev. Lett. **59**, 672–675 (Aug 1987).

[28] W. Ketterle and N. J. Van Druten, "Evaporative cooling of trapped atoms," Advances In Atomic, Molecular and Optical Physics **37**, 181–236 (1996)

[29] F. Dalfovo and S. Giorgini, "Theory of Bose-Einstein condensation in trapped gases," Rev. Mod. Phys. **71**, 463 – 512 (1999)

[30] M. H. Anderson, J. R. Ensher, M. R. Matthews, C. E. Wieman, and E. A. Cornell, "Observation of Bose-Einstein condensation in a gas of sodium atoms," Science **269**, 198–201 (1995)

[31] S. Burger, K. Bongs, S. Dettmer, W. Ertmer, K. Sengstock, A. Sanpera, G. V. Shlyapnikov, and M. Lewenstein, "Dark solitons in Bose-Einstein condensates," Phys. Rev. Lett. **83**, 5198–5201 (Dec 1999)

[32] M. Greiner, O. Mandel, T. W. Hänsch, and I. Bloch, "Collapse and revival of the matter wave field of a BoseEinstein condensate," Nature **419**, 51–54 (2002)

[33] C. Gross, T. Zibold, E. Nicklas, J. Esteve, and M. K. Oberthaler, "Nonlinear atom interferometer surpasses classical precision limit," Nature **464**, 1166–1169 (2010)

[34] M.-O. Mewes, M. R. Andrews, D. M. Kurn, D. S. Durfee, C. G. Townsend, and W. Ketterle, "Output coupler for Bose-Einstein condensed atoms," Phys. Rev. Lett. **78**, 582–585 (Jan 1997)

[35] I. Bloch, T. W. Hänsch, and T. Esslinger, "Atom laser with a cw output coupler," Phys. Rev. Lett. **82**, 3008–3011 (Apr 1999)

[36] J. Fortagh, A. Grossmann, C. Zimmermann, and T. W. Hänsch, "Miniaturized wire trap for neutral atoms," Phys. Rev. Lett. **81**, 5310–5313 (Dec 1998)

[37] M. P. A. Jones, C. J. Vale, D. Sahagun, B. V. Hall, and E. A. Hinds, "Spin coupling between cold atoms and the thermal fluctuations of a metal surface," Phys. Rev. Lett. **91**, 080401 (Aug 2003)

[38] M. Drndić, K. S. Johnson, J. H. Thywissen, M. Prentiss, and R. M. Westervelt, "Microelectromagnets for atom manipulation," Appl. Phys. Lett. **72**, 22 (1998)

[39] J. Fortágh, H. Ott, G. Schlotterbeck, C. Zimmermann, B. Herzog, and D. Wharam, "Microelectromagnets for trapping and manipulating ultracold atomic quantum gases," Appl. Phys. Lett. **81**, 1146–1148 (2002)

[40] J. Estève, C. Aussibal, T. Schumm, C. Figl, D. Mailly, I. Bouchoule, C. I. Westbrook, and A. Aspect, "Role of wire imperfections in micromagnetic traps for atoms," Phys. Rev. A **70**, 043629 (Oct 2004)

[41] S. Groth, P. Krüger, S. Wildermuth, R. Folman, T. Fernholz, J. Schmiedmayer, D. Mahalu, and I. Bar-Joseph, "Atom Chips: Fabrication and thermal properties," Appl. Phys. Lett. **85**, 2980 (2004)

[42] M. Trinker, S. Groth, L. Haslinger, S. Manz, T. Betz, S. Schneider, I. Bar-Joseph, T. Schumm, and J. Schmiedmayer, "Multilayer atom chips for versatile atom micromanipulation," Applied Physics Letters **92**, 254102 (2008)

[43] W. Hänsel, P. Hommelhoff, T. W. Hänsch, and J. Reichel, "Bose-Einstein condensation on a microelectronic chip," Nature **413**, 498–501 (2001)

[44] H. Ott, J. Fortagh, G. Schlotterbeck, A. Grossmann, and C. Zimmermann, "Bose-Einstein condensation in a surface microtrap," Phys. Rev. Lett. **87**, 230401 (Nov 2001)

[45] E. A. Hinds and I. G. Hughes, "Magnetic atom optics: mirrors, guides, traps, and chips for atoms," Journal of Physics D **32**, R119–R146 (1999)

[46] R. Gerritsma, S. Whitlock, T. Fernholz, H. Schlatter, J. A. Luigjes, J.-U. Thiele, J. B. Goedkoop, and R. J. C. Spreeuw, "Lattice of microtraps for ultracold atoms based on patterned magnetic films," Phys. Rev. A **76**, 033408 (Sep 2007)

[47] S. Pollock, J. P. Cotter, A. Laliotis, and E. A. Hinds, "Integrated magneto-optical traps on a chip using silicon pyramid structures," Optics Express **17**, 16 (2009)

[48] M. Kohnen, M. Succo, P. G. Petrov, R. A. Nyman, M. Trupke, and E. A. Hinds, "An integrated atom-photon junction," arXiv **0912.4460v1** (2009)

[49] S. Eriksson, M. Trupke, H. F. Powell, D. Sahagun, C. D. J. Sinclair, E. A. Curtis, B. E. Sauer, E. A. Hinds, Z. Moktadir, C. O. Gollasch, and M. Kraft, "Integrated optical components on atom chips," Eur. Phys. J. D **35**, 135–139 (2005)

[50] M. Trupke, J. Goldwin, B. Darquié, G. Dutier, S. Eriksson, J. Ashmore, and E. A. Hinds, "Atom detection and photon production in a scalable, open, optical microcavity," Phys. Rev. Lett. **99**, 063601 (Aug 2007)

[51] T. P. Purdy and D. M. Stamper-Kurn, "Integrating cavity quantum electrodynamics and ultracold-atom chips with on-chip dielectric mirrors and temperature stabilization," Applied Physics B **90**, 401–405 (2008)

[52] M. Wilzbach, D. Heine, S. Groth, X. Liu, T. Raub, B. Hessmo, and J. Schmiedmayer, "Simple integrated single-atom detector," Optics Letters **34**, 259–261 (2009)

[53] M. R. Andrews, C. G. Townsend, H. J. Miesner, D. S. Durfee, D. M. Kurn, and W. Ketterle, "Observation of interference between two Bose condensates," Science **275**, 637 (1997)

[54] R. P. Feynman and A. R. Hibbs, *Quantum Mechanics and Path Integrals* (McGraw-Hill, 1965)

[55] M. E. Peskin and D. V. Schroeder, *An Introduction to Quantum Field Theory* (Westview Press Inc., 1995)

[56] Y. Aharonov and D. Bohm, "Significance of electromagnetic potentials in the quantum theory," Phys. Rev. **115**, 485–491 (Aug 1959)

[57] E. P. Gross, "Structure of a quantized vortex in boson systems," Il Nuovo Cimento **20**, 454 (1961)

[58] L. P. Pitaevskii, Zh. Eksp. Teor. Fiz. **40**, 646 (1961)

[59] N. Bogoliubov, "On the theory of superfluidity," Journal of Physics **11**, 23–32 (1947)

[60] N. D. Mermin and H. Wagner, "Absence of ferromagnetism or antiferromagnetism in one- or two-dimensional isotropic Heisenberg models," Phys. Rev. Lett. **17**, 1133–1136 (Nov 1966)

[61] P. C. Hohenberg, "Existence of long-range order in one and two dimensions," Phys. Rev. **158**, 383–386 (Jun 1967)

[62] D. S. Petrov, G. V. Shlyapnikov, and J. T. M. Walraven, "Regimes of quantum degeneracy in trapped 1d gases," Phys. Rev. Lett. **85**, 3745–3749 (Oct 2000)

[63] S. Dettmer, D. Hellweg, P. Ryytty, J. J. Arlt, W. Ertmer, K. Sengstock, D. S. Petrov, G. V. Shlyapnikov, H. Kreutzmann, L. Santos, and M. Lewenstein, "Observation of phase fluctuations in elongated Bose-Einstein condensates," Phys. Rev. Lett. **87**, 160406 (Oct 2001)

[64] Y. Castin and J. Dalibard, "Relative phase of two Bose-Einstein condensates," Phys. Rev. A **55**, 4330–4337 (Jun 1997)

[65] J. Javanainen and S. M. Yoo, "Quantum phase of a Bose-Einstein condensate with an arbitrary number of atoms," Phys. Rev. Lett. **76**, 161–164 (Jan 1996)

[66] L. Pitaevskii and S. Stringari, *Bose-Einstein condensation* (Clarendon Press, 2003)

[67] B. J. Dalton, "Two-mode theory of BEC interferometry," Journal of Modern Optics **54**, 615 (2007)

[68] J. Javanainen and M. Y. Ivanov, "Splitting a trap containing a Bose-Einstein condensate: atom number fluctuations," Phys. Rev. A **60**, 2351–2359 (Sep 1999)

[69] M. O. Scully and M. S. Zubairy, *Quantum Optics* (Cambridge University Press, 1997)

[70] L. Pezze, A. Smerzi, G. P. Berman, A. R. Bishop, and L. A. Collins, "Dephasing and breakdown of adiabaticity in the splitting of Bose-Einstein condensates," New Journal of Physics **7**, 85 (2005)

[71] J. Javanainen and M. Wilkens, "Phase and phase diffusion of a split Bose-Einstein condensate," Phys. Rev. Lett. **78**, 4675–4678 (Jun 1997)

[72] G. Baym and C. J. Pethick, "Ground-state properties of magnetically trapped Bose-Condensed rubidium gas," Phys. Rev. Lett. **76**, 6–9 (Jan 1996)

[73] Nicholas K. Whitlock and Isabelle Bouchoule, "Relative phase fluctuations of two coupled one-dimensional condensates," Phys. Rev. A **68**, 053609 (Nov 2003)

[74] A. Polkovnikov, E. Altman, and E. Demler, "Interference between independent fluctuating condensates," Proc. Natl. Acad. Sci. USA **103**, 6125 (2006)

[75] A. Vogel, M. Schmiedt, K. Sengstock, K. Bongs, W. Lewoczko, T. Schuldt, A. Peters, T. Van Zoest, W. Ertmer, E. Rasel, T. Steinmetz, J. Reichel, T. Koenemann, W. Brinkmann, E. Goeklue, C. Laemmerzahl, H. J. Dittus, G. Nandi, W. P. Schleich, and R. Walser, "Bose-Einstein condensates in microgravity," Applied Physics B **84**, 4 (2006)

[76] M. Jones, *Bose-Einstein condensation on an atom chip*, Ph.D. thesis, Imperial College London (2003)

[77] D. S. Sanchez, *Cold atoms trapped near surfaces*, Ph.D. thesis, Imperial College London (2006)

[78] R. J. Sewell, *Matter wave interference on an atom chip*, Ph.D. thesis, Imperial College London (2009)

[79] C. D. J. Sinclair, *Bose-Einstein condensation in microtraps on videotape*, Ph.D. thesis, University of London (2005)

[80] R.J.Sewell, J. Dingjan, F. Baumgärtner, I. Llorente-Garcia, S. Eriksson, E. A. Hinds, G. Lewis, P. Srinivasan, Z. Moktadir, C. O. Gollash, and M. Kraft, "Atom chip for BEC interferometry," J. Phys. B: At. Mol. Opt. Phys. **43**, 051003 (2010)

[81] A. Arnold, J. Wilson, and M. Boshier, "A simple extended-cavity diode laser," Review of Scientific Instruments **69**, 1236 (1998)

[82] W. Ketterle, D.S. Durfree, and D. M. Stamper-Kurn, "Making, probing and understanding Bose-Einstein condensates," in *Proceedings of the 1998 Enrico Fermi school on Bose-Einstein condensation in Varenna, Italy* (Academic Press, 1998)

[83] P. Horowitz and W. Hill, *The Art of Electronics* (Cambridge University Press, 1989)

[84] M. Succo and C. Ospelkaus, "VFG-Control Project Homepage," http://vfg-control.sourceforge.net/index.html

[85] J. Reichel, W. Hänsel, and T. W. Hänsch, "Atomic micromanipulation with magnetic surface traps," Phys. Rev. Lett. **83**, 3398–3401 (Oct 1999)

[86] Y. V. Gott, M. S. Ioffe, and V. G. Telkovskii, Nuclear Fusion Suppl. Pt. **3**, 1045 (1962)

[87] D. E. Pritchard, "Cooling neutral atoms in a magnetic trap for precision spectroscopy," Phys. Rev. Lett. **51**, 1336–1339 (Oct 1983)

[88] V. S. Bagnato, G. P. Lafyatis, A. G. Martin, E. L. Raab, R. N. Ahmad-Bitar, and D. E. Pritchard, "Continuous stopping and trapping of neutral atoms," Phys. Rev. Lett. **58**, 2194–2197 (May 1987)

[89] J. Reichel, "Microchip traps and BoseEinstein condensation," Applied Physics B: Lasers and Optics **75**, 469–487 (2002)

[90] J. Fortágh, H. Ott, S. Kraft, A. Günther, and C. Zimmermann, "Surface effects in magnetic microtraps," Phys. Rev. A **66**, 041604 (Oct 2002)

[91] A. E. Leanhardt, Y. Shin, A. P. Chikkatur, D. Kielpinski, W. Ketterle, and D. E. Pritchard, "Bose-Einstein condensates near a microfabricated surface," Phys. Rev. Lett. **90**, 100404 (Mar 2003)

[92] M. P. A. Jones, C. J. Vale, D. Sahagun, B. V. Hall, C. C. Eberlein, B. E. Sauer, K. Furusawa, D. Richardson, and E. A. Hinds, "Cold atoms probe the magnetic field near a wire," Journal of Physics B: At. Mol. Opt. Phys. **37**, L15–L20 (2004)

[93] P. Krüger, L. M. Andersson, S. Wildermuth, S. Hofferberth, E. Haller, S. Aigner, S. Groth, I. Bar-Joseph, and J. Schmiedmayer, "Potential roughness near lithographically fabricated atom chips," Phys. Rev. A **76**, 063621 (Dec 2007)

[94] D.-W. Wang, M. D. Lukin, and E. Demler, "Disordered Bose-Einstein condensates in quasi-one-dimensional magnetic microtraps," Phys. Rev. Lett. **92**, 076802 (Feb 2004)

[95] W. Petrich, M. H. Anderson, J. R. Ensher, and E. A. Cornell, "Stable, tightly confining magnetic trap for evaporative cooling of neutral atoms," Phys. Rev. Lett. **74**, 3352–3355 (Apr 1995)

[96] P. W. Milonni and J. H. Eberly, *Lasers* (John Wiley & Sons, 1988)

[97] Sinclair Optics Inc., "OSLO," http://www.sinopt.com

[98] J. J. P. van Es, S. Whitlock, T. Fernholz, A. H. van Amerongen, and N. J. van Druten, "Longitudinal character of atom-chip-based rf-dressed potentials," Phys. Rev. A **77**, 063623 (Jun 2008)

[99] S. Hofferberth, I. Lesanovsky, B. Fischer, T. Schumm, and J. Schmiedmayer, "Non-equilibrium coherence dynamics in one-dimensional Bose gases," Nature **449**, 324 (2007)

[100] W. Demtröder, *An Introduction to Atomic and Molecular Physics* (Springer, 2005)

[101] D. E. Pritchard, K. Helmerson, and A. G. Martin, *Atom traps* (World Scientific, 1989)

[102] K. B. Davis, M. O. Mewes, M. R. Andrews, N. J. van Druten, D. S. Durfee, D. M. Kurn, and W. Ketterle, "Bose-Einstein condensation in a gas of sodium atoms," Phys. Rev. Lett. **75**, 3969–3973 (Nov 1995)

[103] C. C. Bradley, C. A. Sackett, J. J. Tollett, and R. G. Hulet, "Evidence of Bose-Einstein condensation in an atomic gas with attractive interactions," Phys. Rev. Lett. **75**, 1687–1690 (Aug 1995)

[104] J. Dalibard and C. N. Cohen-Tannoudji, "Dressed-atom approach to atomic motion in laser light: the dipole force revisited," Journal of Optical Society America **B2**, 1707 (1985)

[105] C. N. Cohen-Tannoudji, "Manipulating atoms with photons," Rev. Mod. Phys. **70**, 707–741 (1998)

[106] S. Hofferberth, B. Fischer, T. Schumm, J. Schmiedmayer, and I. Lesanovsky, "Ultracold atoms in radio-frequency dressed potentials beyond the rotating-wave approximation," Phys. Rev. A **76**, 013401 (2007)

[107] I. I. Rabi, N. F. Ramsey, and J. Schwinger, "Use of rotating coordinates in magnetic resonance problems," Rev. Mod. Phys. **26**, 167–171 (Apr 1954)

[108] C. L. Garrido Alzar, H. Perrin, B. M. Garraway, and V. Lorent, "Evaporative cooling in a radio-frequency trap," Phys. Rev. A **74**, 053413 (Nov 2006)

[109] R. K. Easwaran, L. Longchambon, P.-E. Pottie, V. Lorent, H. Perrin, and B. M. Garraway, "Rf spectroscopy in a resonant rf-dressed trap," J. Phys. B: At. Mol. Opt. Phys. **43**, 065302 (2010)

[110] S. Haroche, C. Cohen-Tannoudji, C. Audoin, and J. P. Schermann, "Modified Zeeman hyperfine spectra observed in $H1$ and $Rb87$ ground states interacting with a non-resonant rf field," Phys. Rev. Lett. **24**, 861–864 (Apr 1970)

[111] I. Lesanovsky, T. Schumm, S. Hofferberth, L. M. Andersson, P. Krüger, and J. Schmiedmayer, "Adiabatic radio-frequency potentials for the coherent manipulation of matter waves," Phys. Rev. A **73**, 033619 (Mar 2006)

[112] T. Schumm, *Bose-Einstein condensates in magnetic double well potentials*, Ph.D. thesis, University of Heidelberg (2005)

[113] B. V. Hall, S. Whitlock, R. Anderson, P. Hannaford, and A. I. Sidorov, "Condensate splitting in an asymmetric double well for atom chip based sensors," Phys. Rev. Lett. **98**, 030402 (Jan 2007)

[114] J. Estève, C. Gross, A. Weller, S. Giovanazzi, and M. K. Oberthaler, "Squeezing and entanglement in a Bose-Einstein condensate," Nature **455**, 1216–1219 (2008)

[115] A. Roehrl, M. Naraschewski, A. Schenzle, and H. Wallis, "Transition from phase locking to the interference of independent Bose condensates: theory versus experiment," Phys. Rev. Lett. **78**, 4143–4146 (Jun 1997)

[116] S. Raghavan, A. Smerzi, S. Fantoni, and S. R. Shenoy, "Coherent oscillations between two weakly coupled Bose-Einstein condensates: Josephson effects, π oscillations, and macroscopic quantum self-trapping," Phys. Rev. A **59**, 620–633 (Jan 1999)

[117] T. Schumm, P. Krüger, S. Hofferberth, I. Lesanovsky, S. Wildermuth, S. Groth, I. Bar-Joseph, L. M. Andersson, and J. Schmiedmayer, "A double well interferometer on an atom chip," Quantum Information Processing **5**, 537–558 (2006)

[118] J. J. P. van Es, *Bose-Einstein condensates in radio-frequency-dressed potentials on an atom chip*, Ph.D. thesis, University Amsterdam (2009)

[119] S. Hofferberth, I. Lesanovsky, T. Schumm, J. Schmiedmayer, A. Imambekov, V. Gritsev, and E. Demler, "Probing quantum and thermal noise in an interacting many-body system," Nature Physics **4**, 489–495 (2008)

[120] K. V. Mardia and P. E. Jupp, *Directional Statistics* (John Wiley & Sons, 2000)

[121] R. Scott, "Private Communication,"

[122] B. D. Josephson, "Possible new effects in superconducting tunnelling," Phys. Lett. **1**, 251 (1962)

[123] A. Smerzi, S. Fantoni, S. Giovanazzi, and S. R. Shenoy, "Quantum coherent atomic tunneling between two trapped Bose-Einstein condensates," Phys. Rev. Lett. **79**, 25 (1997)

[124] M. Albiez, R. Gati, J. Fölling, S. Hunsmann, M. Cristiani, and M. K. Oberthaler, "Direct observation of tunneling and nonlinear self-trapping in a single Bosonic Josephson junction," Phys. Rev. Lett. **95**, 010402 (Jun 2005)

[125] C. Menotti, J. R. Anglin, J. I. Cirac, and P. Zoller, "Dynamic splitting of a Bose-Einstein condensate," Phys. Rev. A **63**, 023601 (Jan 2001)

[126] M. Fattori, C. D'Errico, G. Roati, M. Zaccanti, M. Jona-Lasinio, M. Modugno, M. Inguscio, and G. Modugno, "Atom interferometry with a weakly interacting Bose-Einstein condensate," Phys. Rev. Lett. **100**, 080405 (Feb 2008)

[127] E. A. Hinds and V. Sandoghdar, "Cavity QED level shifts of simple atoms," Phys. Rev. A **43**, 398–403 (Jan 1991)

[128] C. I. Sukenik, M. G. Boshier, D. Cho, V. Sandoghdar, and E. A. Hinds, "Measurement of the Casimir-Polder force," Phys. Rev. Lett. **70**, 560–563 (Feb 1993)

[129] D. M. Harber, J. M. Obrecht, J. M. McGuirk, and E. A. Cornell, "Measurement of the Casimir-Polder force through center-of-mass oscillations of a Bose-Einstein condensate," Phys. Rev. A **72**, 033610 (Sep 2005)

[130] J. M. Obrecht, R. J. Wild, M. Antezza, L. P. Pitaevskii, S. Stringari, and E. A. Cornell, "Measurement of the temperature dependence of the Casimir-Polder force," Phys. Rev. Lett. **98**, 063201 (Feb 2007)

[131] G. Bressi, G. Carugno, R. Onofrio, and G. Ruoso, "Measurement of the Casimir force between parallel metallic surfaces," Phys. Rev. Lett. **88**, 041804 (Jan 2002)

[132] S. K. Lamoreaux, "Demonstration of the Casimir force in the 0.6 to 6 μm range," Phys. Rev. Lett. **78**, 5–8 (Jan 1997)

I want morebooks!

Buy your books fast and straightforward online - at one of world's fastest growing online book stores! Environmentally sound due to Print-on-Demand technologies.

Buy your books online at
www.morebooks.shop

Kaufen Sie Ihre Bücher schnell und unkompliziert online – auf einer der am schnellsten wachsenden Buchhandelsplattformen weltweit! Dank Print-On-Demand umwelt- und ressourcenschonend produziert.

Bücher schneller online kaufen
www.morebooks.shop

info@omniscriptum.com
www.omniscriptum.com

Printed by Books on Demand GmbH, Norderstedt / Germany